ROAD AND BRIDGE CONSTRUCTION HANDBOOK

ROAD AND BRIDGE CONSTRUCTION HANDBOOK

Michael Lapinski

VAN NOSTRAND REINHOLD COMPANY
NEW YORK CINCINNATI ATLANTA DALLAS SAN FRANCISCO
LONDON TORONTO MELBOURNE

Van Nostrand Reinhold Company Regional Offices:
New York Cincinnati Atlanta Dallas San Francisco

Van Nostrand Reinhold Company International Offices:
London Toronto Melbourne

Library of Congress Catalog Card Number: 78-1475
ISBN: 0-442-24681-1

Manufactured in the United States of America

Published by Van Nostrand Reinhold Company
135 West 50th Street, New York, N. Y. 10020

Published simultaneously in Canada by Van Nostrand Reinhold Ltd.

15 14 13 12 11 10 9 8 7 6 5 4 3 2 1

Library of Congress Cataloging in Publication Data

Lapinski, Michael.
Road and bridge construction handbook.

Includes index.
1. Road construction. 2. Bridge construction.
I. Title.
TE153.L32 625.7 78-1475
ISBN 0-442-24681-1

Introduction

Heavy construction today is a complex technology. Gone are the days when roads and bridges were constructed by "eyeballing" or "seat of the pants" guesswork. Such methods sounded simple, and the method of pay was often just as roughly calculated. Today there are plans and specifications to which all roads and structures must be built. They require a level of precision and quality that often baffles the construction worker.

Even the most simple construction jobs have changed to achieve optimum results. Today timber haul roads and gravel county roads are built to almost the same identical specifications as a four-lane freeway. Gone are the days when the small dirt road cost 30¢ per cubic yard. Today the cost per cubic yard for a dirt road is almost as much as for a freeway job.

These converging standards of quality have brought about changes among contractors. It used to be that the big freeway projects were the sole domain of the larger construction firms, and smaller contractors were given the less lucrative dirt road projects. But today, there are more small firms bidding on freeways, and more of the larger firms are bidding on county roads. Why? Because of the uniformity of standards and pay.

Nowhere has this change been more difficult and challenging than in actual construction practices. Old methods must be cast aside and special know how and experience developed to deliver a product of such high standards. The problem is not only delivery, but making money at the same time. For a job to be profitable today, every phase of construction from earthwork, to maintenance, to supervision must function together. Supervisors, foremen, and workers must all have the same basic working knowledge of each phase of construction, and must be able to determine the most economical method for each situation.

The purpose of this book is to explain the construction methods for roads and structures in simple everyday terms so that the construction worker will have a good basic knowledge of his occupation. Each phase of construction is broken down to the how and why so that the man standing in the fresh damp dirt will better understand his job and choose the most economical approach. This book also delves into the critical bidding stage, when profits are often lost even before a project is undertaken due to insufficient knowledge of construction costs and technical requirements.

Chapter 17, entitled "The Supervision And Orderly Operation of a Construction Project" transcends construction machinery and methods. It discusses the basic management of men—the single most valuable resource on any job. How to direct, manipulate, demand respect, and obtain optimum results from employees has long been one of those intangibles of the construction industry.

This book is also intended for another member of the construction team—the civil engineer. Engineers, especially those just out of college, are crammed with analytical data and classroom examples of construction problems. No wonder they are bewildered when they enter the construction industry and discover that much of their schooling has no immediate application to the fast-paced construction industry. Many will become designers with only vague ideas of how their designs adapt to a project in a practical sense. This book translates much of the high education of the civil engineer into basic common sense principles.

Contents

ROAD AND BRIDGE CONSTRUCTION HANDBOOK

1
Bidding A Job

The bidding stage of a project is so critical that most large construction firms use only their most experienced and competent people as bidders. These employees do nothing else but bid on prospective contracts. They must know the best way to perform each item of work, and the costs involved. Costs are made up of a number of factors such as the direct costs for labor and materials, plus the costs of doing business, called *overhead*. To the total is added a profit margin, usually 16%. Accuracy and good judgment are vital in preparing a bid. For example, on a road job with 200,000 cubic yards of excavation, a one cent variation in cost per yard makes a difference of $2000.

A construction project is advertised, usually in local and regional newspapers and trade papers, for a minimum of 30 days. Fact and bid sheets can be obtained from the contracting office where the bid originated. After poring over the work items and specifications, and carefully preparing a cost estimate, a contractor submits his sealed bid to the contracting office to await the bid opening date. If all bids are found to exceed the engineer's cost estimate, the contract may not be awarded. Instead, it is readvertised in hopes of getting lower bids. If these bids are still high, it is assumed that an error has been made in the design engineer's estimate, and the contract is then awarded to the lowest bidder.

There are four major types of construction contracts. The most common is the one in which a contractor bids on and is paid for each item of work such as structural excavation, clearing, etc. Most roads and bridges fall into this category. The second type is used when a contractor is paid for the weight of the material produced. Most riprap and gravel jobs are of this type. The third is for equipment rental. A contractor bids his equipment at an hourly rate to be operated by his personnel under the direction of the engineer. Maintenance, slide removal, and small riprap jobs often are equipment rental contracts. The fourth is a lump sum contract. For this type a contractor looks over a job with no designed quantities to help estimate the cost. A total amount, or lump sum bid is then submitted, with no provision for extra compensation for added work. Contractors should limit their bids to the first three types of contracts. The lump sum bid is always for the work of private landowners and should be avoided because of the lack of designed quantities. In other words, a contractor has no way of knowing how much extra work he might have to do without compensation.

BIDDING ROAD CONSTRUCTION

Most road projects are covered by "designed items of work" contracts, which are the fairest and safest for both parties. A road is designed, and the calculated quantities of earthwork, clearing, seeding, culverts, pavement, etc., are entered on a bid sheet (Fig. 1-1). When a contract is advertised any interested party can obtain the bid sheet from the contracting office.

Each item of work must be bid on individually. The entire contract, however, is awarded to the lowest bid for all items combined. These individual bids are very important because all quantity overruns are paid for at the bid rate. For example, an overrun of 5000 cubic yards of excavation will be paid at the rate which in Fig. 1-1 is $1.50 per cubic yard. The importance of itemized bids cannot be overemphasized. The contractor must be able to live with these bids at the time of contract award and during construction. Designed quantities change during the course of a job. If an item is

Fig. 1-1. Typical bid sheet for a construction project. The underlined areas are blank when issued to a contractor and must be filled in and signed by a contracting representative.

ROAD PROJECT NO. 302.6
STATE HIGHWAY 52
5.2 MILES CONSTRUCTION

ITEM OF WORK	ESTIMATED QUANTITY	UNIT BID	TOTAL
UNCLASSIFIED EXCAVATION	200,000 CU YDS	$ 1.50	$300,000
18 IN. DIA. CULVERT PIPE	560 LIN. FT	12.32	6,899
36 IN. DIA. CULVERT PIPE	216 LIN. FT	15.20	3,283
STRUCTURAL EXCAVATION	220 CU YDS	9.60	2,112
CLEARING MEDIUM	36.7 ACRES	2100	77,070
CLEARING, HEAVY	23.1 ACRES	2600	60,060
OVERHAUL	33,689 STATION YDS	0.23	7,748
RIPRAP	240 CU YDS	6.90	1,656
GUARDRAIL	230 LIN. FT	14.00	3,200
HOT ASPHALT PAVEMENT (BASE COURSE)	25,187 TONS	13.88	349,595
HOT ASPHALT PAVEMENT (SURFACE COURSE)	4,904 TONS	16.95	83,122
FERTILIZE AND SEED	31.3 ACRES	360	11,268
DELINEATORS	1850 PIECES	4.35	8,047
		GRAND TOTAL	$914,080

CONTRACTING FIRM: B & J CONSTRUCTION INC.

CONTRACTING REPRESENTATIVE ________________ DATE ________

bid low to win the contract and the quantity of material changes, a serious loss could occur.

A road contract should never be bid without first walking the job. Many intangibles exist that do not show up in the contract advertisement. For example, excavation and clearing on steep terrain will be more costly than flat or gently sloping land. Culvert lengths cannot be ignored. There is an appreciable cost difference for pipes laid in steep, narrow draws where equipment cannot reach the work area. The installation of simple cross drains, on the other hand, is straightforward. When culverts are included on a bid sheet, always make sure there is compensation for structural excavation. This is the excavation required to bed the pipe to the staked depth. Sometimes this quantity is incorporated under unclassified excavation, but to play safe, question the contracting office if it is not shown as a bid item. The cost for structural excavation is about eight times more than for unclassified excavation.

Most contracts call for unclassified excavation. This means that there is no guarantee what percentage will be common dirt and what will be rock. In other words, it is up to the contractor to look over the project site, determine the amount of rock, and increase his bid accordingly. The cost ratio of common dirt vs rock is roughly as follows: If dirt is \$1.00 per cubic yard, then rippable rock (which can be torn loose with a dozer) should be \$2.00; and hard rock (which must be drilled) should be \$5.00. Contractors have lost thousands of dollars thinking that the term "unclassified excavation" meant common dirt.

How does a contractor determine if there is rock before the ground is opened up? There are several indicators. The most obvious sign of rock is in rock outcrops. To be safe, figure at least a hundred feet of hard rock on either side. Another less apparent sign of hidden rock is in ridges. If the road winds in and out through draws and around ridge points, allow for some rock. Ridges are formed because they resist erosion due to rock under the surface.

Steep slopes with open grassy areas indicate a lack of soil depth, and that rock is close to the surface. If trees in an area are smaller than normal and lean downhill, this too indicates a lack of soil depth. When tree roots are forced to the surface, this indicates rock is close to the surface and will not let the roots penetrate. Where

APPROXIMATE COSTS FOR ROAD CONSTRUCTION AS OF JULY 1978

WORK ITEM	COST (IN DOLLARS)
EXCAVATION:	
COMMON	0.90 PER CU YD
RIPPABLE ROCK	2.50 PER CU YD
SOLID ROCK	4.00 PER CU YD
CULVERT PIPE:	
18 IN. DIA.	12.00 PER LIN. FT[a]
24 IN. DIA.	16.00 PER LIN. FT
36 IN. DIA.	20.00 PER LIN. FT
48 IN. DIA.	29.00 PER LIN. FT
CLEARING:	
LIGHT	1,500 PER ACRE
MEDIUM	2,100 PER ACRE
HEAVY	2,700 PER ACRE
OVERHAUL	0.10–0.20 PER STATION YD
SEED AND FERTILIZE	250 PER ACRE[b]
BITUMINOUS PRIME COAT	0.06 PER SQ YD
HOT MIX ASPHALT BASE COURSE	12.00 PER TON
HOT MIX ASPHALT SURFACE COURSE	14.00 PER TON
GRAVEL (1 IN. MINUS)	8.50 PER TON
CONCRETE	35.00 PER CU YD

NOTE: ADD 9% TO COSTS FOR INFLATION FOR SUCCEEDING YEARS.

[a] FURNISHED AND INSTALLED.
[b] FOR MULCHING ADD $100 PER ACRE.

Fig. 1-2. Approximate costs for road construction. When bidding unfamiliar items, contact a local contractor for local cost deviations.

such signs of surface rock exist, a safe estimate is about 60% solid rock and 40% overburden and rippable rock.

Often, a contract has items of work that a contractor is not set up to handle, such as pavements or steel bridges. Some contractors are frightened away because they don't have the equipment or know how. The smart road building contractor will consult a paving or bridge contractor to handle such special projects for him, get their price and bid the job. He is called the *prime contractor.* Placing parts of the work with other firms is called *subcontracting*. Some prime contractors have as many as six subcontractors on a project at the same time.

Figure 1-2 gives approximate costs for both dirt road and freeway road construction. This shows a national average. Each locality varies somewhat, and a contractor should check with other local contractors before submitting a bid on any unfamiliar items.

RIPRAP AND GRAVEL JOBS

These contracts are usually bid by the measured weight of the material created. The weight per yard of material is determined through direct testing, and this weight per yard of rock or gravel is the basis of payment for the work. Large certified truck scales, usually furnished by the contractor, weigh each truckload of material before it is used. This is the most satisfactory and fairest method of payment, and therefore, a simple item to bid.

While rock quantities comprise most of the compensation, there are several other items of work that are costly and for which payment should be received. On riprap jobs, keyways and sloughed-out banks (see Chapter 13, Fig. 13.1) should be either bid separately or incorporated into the price of the rock if they are subsidiary items. Keyways are especially expensive to construct. Once riprap has been dumped into a work area, it must be smoothed or combed to a reasonably smooth surface by a crane or some other machine. This is usually a subsidiary item, but should be included in the overall riprap costs. On gravel jobs, the new and old roadway must be bladed smooth and ditches cleaned with graders. Also, all protruding rocks must be either removed or blasted. This cost is about $400 per mile. Contracts usually include this work as a bid item, otherwise compensate for it in the gravel cost.

Once gravel is dumped on the road, it must then be processed by mixing the segregated large and small crushed rock (See Chapter 10). This costs about $150 per mile unless the gravel must be watered to obtain the required compaction (which it usually does). If so, the cost will increase another $50 per mile. Processing is rarely a bid item so include it as part of the bid price for the gravel.

For the riprap and gravel contracts, the quarry can be the source of countless headaches. Quarries are analyzed by geologists from the contracting office, and a copy of their findings should be available to all bidders. Things to look for include extra hard rock that will be more expensive to break because of additional drill holes and powder. If large dirt seams are evident in the rock, a contractor may be faced with rejection of his riprap or gravel because of too much dirt. If a rock source looks troublesome while walking the job, and the geologist's report confirms this, it may be wise to pass up the contract.

Clearing of the quarry area and its rehabilitation after the rock has been removed should also be bid items. If not, ask the contracting engineer about compensation for them.

BRIDGES AND RETAINING WALLS

Bridges and retaining walls, like roads, are bid using designed estimated items of work. Some of the items on a bridge job include structural excavation, reinforcement steel, concrete, guardrails, prefabricated stringers, etc. Bid items on a retaining wall include structural excavation, wall material (steel, concrete, or timbers), backfill, and guardrails.

Contracts for prefabricated bridges or retaining walls may stipulate that the contractor furnish the structure. This requirement should not deter a prospective bidder since such structures can be ordered from local bridge or steel firms. All that the contractor must do is furnish them with a set of standard drawings which can be obtained from the contracting office. Normally, a prefabricated structure is shown on the bid sheet as a single item. When bidding on prefabricated structures, first call one or more firms to get manufactured and shipping costs. The contractor then adds his 16% profit and an amount to cover his labor cost for erecting it. The result will be an accurate and safe bid.

Individual bid items are critical in structural contracts. The bid on concrete may run as high as $300 per cubic yard, and structural excavation $20 per cubic yard. It is obvious that slight increases in their quantities can significantly increase the contractor's return.

Labor plays a large part in structure work. Manpower is needed to construct plywood concrete forms and tie reinforcement bars on bridge jobs. Weeks are spent putting together a retaining wall structure. That is why certain bid items such as concrete are priced so high—in order to compensate for the enormous cost of labor. In order to apply the correct cost of labor to a job, a contractor must consult with his foreman to determine first, how much work can be accomplished for each dollar spent, and how many men it will take for maximum efficiency. From this basic data a fairly accurate labor cost can be determined for each work phase. Each major work item to be bid is made up of one or more work phases. These costs are applied to the work item as required.

Another factor that frightens many prospective bidders is the complexity of the work. Somehow word often gets around that a contractor who builds dirt roads cannot build a bridge because the work is too complicated. This is not true. With a set of contract drawings and the basic know how supplied in Chapter 14, any contractor should be able to bid and construct a bridge structure.

2
Contract Specifications

Contract specifications are the rules that the man on the job must deal with. They direct and regulate the quality of each item of work. Specifications cannot be ignored by the man who bids the job. For instance, excavation which specifies that the fills be built in one foot layers and compacted to 95% optimum density is more expensive than a sidecast fill. However, after the bid is won, it is the man on the job who must deal with the "specs" every day of the contract.

This chapter will cover the major specifications that a contractor will encounter, point out areas to watch out for, and explain their intent.

ROADWAY EXCAVATION

Material from roadway excavation is usually removed by a bulldozer. The specifications say very little about this work other than a reference to the tolerance of the cut slopes. The specifications are more specific for embankments stating whether a contractor can use the easier sidecast method or the more expensive layer placement method.

Sidecast embankment is the simplest method of disposal because the bulldozer simply pushes the material from the cut over the side to the fill stakes. However, with layer placement the embankment material must be laid in maximum one-foot layers until the fill is to grade level. This requires the dozer building the road down the embankment to the bottom of the fill area and pushing the dirt down there in one-foot deep layers.

The most restrictive earthwork specification calls for layer-placed fills compacted to 95% of maximum density, with mandatory compaction required by drum roller, grid roller, or other means. This is an expensive way to build a road, and is typical of highway jobs.

One trouble area to watch in the specification is the term *unsuitable material* which when encountered in both cut and fill must be removed as directed by the engineer. These could be swampy fill areas or cut areas where springs have turned the material to mush. The contractor is bound by the specification to remove this material, but he must be compensated for it. For example, if 500 cubic yards of muck must be removed from a boggy fill area before good fill material is put in, the contractor must be paid for removing that unsuitable material.

Also when dealing with marshy fill areas, the earthwork specification requires that the embankment area must be of suitable stability to support earthwork equipment. What this means of course, is that in a marshy area a dozer would sink out of sight if it tried to lay a one-foot layer of fill across it as would be required for non-marshy areas. For marshy areas, the fill may be two or even three feet deep to support trucks or a bulldozer.

The pioneering of an embankment area is accomplished by a road which the dozer builds to the bottom of the embankment in order to bring up the fill in one-foot layers. It is impossible to get right down to the fill stakes because the dozer must have a flat road on which to move back and forth. In the process of cutting out this road, the dozer spills dirt ahead until that dirt reaches the fill stakes. Sometimes the dozer road is still several feet above the fill bottom when the dirt has been pushed out to the stakes. The engineer cannot make the contractor begin his one-foot layers any lower than the elevation of that road when the dirt hits the fill stakes. This is always a problem area, with the engineer complaining that the dozer should get closer to the fill stakes with the pioneer road, and the contractor answering that it is impossible. The contractor is right, and the specification backs him up by stating that, "All earthwork equipment must have a suitable base to work on." This means a dozer doesn't have to dive over the side without a road to get down to the very bottom of the fill area.

CLEARING SPECIFICATIONS

Most clearing specs call for complete disposal of unmerchantable trees, snags, brush, and stumps by burning. However, even the toughest specification allows some leeway with the small branches and twigs by stating that all limbs, roots, and sticks 3 inches or less in diameter but not more than 4 feet long may be left if they are scattered. This saves the contractor a lot of raking and countless dollars in "stick picking."

Very often, a contractor does a few jobs in which the engineer overlooks some larger logs and isolated stumps. This is money in the contractor's pocket. Based on these experiences, the contractor mistakenly gets the idea that all engineers will be so understanding, and he bids his jobs with this in mind. Unfortunately, there are many engineers who inspect the clearing to the exact letter of the specifications. A contractor should always bid clearing based on picking up every unspecified piece of wood. If he is allowed to leave a few larger pieces, fine, but if not, at least his bid includes an appropriate price to do the work.

CULVERT SPECIFICATIONS

More disagreement is caused over culvert installations than any other phase of construction. The reason is because this is a slow, costly operation involving many detailed specifications.

One troublesome specification calls for the culvert to be backfilled by machine compaction which cannot exceed 6 inches per layer with no rocks in the fill. It is next to impossible to get all the small rocks out of some soil. Usually, the engineer allows up to 2 inch rocks in the fill. Otherwise the contractor must truck in sand or some other fine soil—a costly measure. If that happens, demand compensation for this extraordinary work. Since bedding sand was not included as a bid item the contractor had the right to assume that the native soil was suitable.

Most specs call for unsuitable bedding material to be removed and replaced with suitable bedding. Muddy creek bottoms and rock beds fall into this category. This removal and replacement delays the work of the culvert crew, sometimes all day. If the contractor must drill and shoot rocks in order to bed a culvert, he should demand reasonable compensation for this extra work.

It is usually the contractor's responsibility to furnish all culvert pipe on a project. When the pipe is delivered to the project, the engineer always demands a certificate of compliance to show that the pipe is manufactured to federal standards. It often happens that the delivery truck driver forgets this certification and the engineer will not allow the pipes to be used until one is obtained. When ordering culverts, remind the company not to forget the compliance certificate.

Culverts are manufactured in 4 foot lengths and connected by overlapping and riveting. The specification requires that culverts shall be laid with the outside circumferental lap pointing upstream and the longitudinal lap on the sides. It takes no extra work to install a culvert that way, but a lackadaisical crew often overlooks this small requirement and the result is rejection of the pipe. The contractor must then dig up the pipe with care to prevent damage, and then simply switch ends. This is a costly mistake and should be watched for constantly.

RIPRAP SPECIFICATIONS

Riprap is ordinarily a low-problem item when contracted for payment by weight. There are, however, problems that can appear in the course of a project. A riprap specification may call for riprap foundation trenches (keyways) to be constructed to a depth of 3 feet below a stream bed. There is an equitable payment for this structural excavation, so the contractor begins to dig the keyways and promptly hits bedrock half a foot down. What recourse does he have? None unless he can get the engineer to write a notice stipulating the keyway is to be dug 3 feet deep <u>or</u> to bedrock. Bedrock, of course, is the most stable foundation for riprap.

Another innocent appearing spec (and this one might be hidden in the environmental impact statement) states, "Riprap shall be placed in such a manner as to not destroy existing vegetation on fill slopes above the eroded areas." What this means is that the riprap may not be end-dumped over the shoulder. The only alternative to a bewildered contractor is to use the exorbitantly expensive derrick placement method which involves individually placing rock with a crane.

Aggregate for road gravel or asphalt mix is a continual headache to keep within the specified gradation. Much of the problem originates with the gravel pit. If a pit contains too much dirt, a contractor will have problems keeping the smaller sizes of gravel within specification. The gravel specification calls for a gradation of sizes that range from 1 inch to less than one-quarter inch. If too much dirt is present a test of a gravel sample will show an excessive amount of small sized particles. If the pit is one huge chunk of solid rock, the contractor will have problems in finding enough fines. Fines are used to bind the crushed rock together in the roadbed and therefore, are a vital ingredient of the gradation. However, a gradation that calls for fines to be in the 8 to 12% range can overlap the range from a low of 7% to a high of 14% without hurting the structural strength of the gravel. A contractor should be in constant contact with the engineer when gravel is borderline so that the entire shift of crushed rock is not rejected by some unknowledgeable inspector.

TEMPERATURE AND WEATHER SPECIFICATIONS

Items such as hot mix asphalt pavement and concrete are relatively trouble-free of problems arising from the specification, except for temperature controls. Asphalt work is cut off at the 60–65° F-range, and concrete work is cut off when the temperature falls to about 40° F.

When jobs are contracted in high country where temperatures turn cold early in the fall, a contractor should take temperature limitations into account. A simple way to find out about conditions in an area is to call the local weather bureau.

Rain does not harm concrete. Actually it is beneficial. However, rain is extremely harmful to an asphalt pavement and will hold up work for the duration of the precipitation plus the time it takes to dry out the subgrade. Some coastal and mountainous areas are known for their September rainy season which may begin in August and run through October. For such areas it is a good idea to build the road one season and pave it the next summer to avoid fighting the weather.

3 Constructing The Pioneer Road

A pioneer road is a rough temporary trail built along the route of a project for the purpose of giving the large construction equipment access to the work. There are three types of pioneer roads: the main right-of-way pioneer to give access to the job, the cut pioneer which gives access to the top of the cut slopes, and the fill pioneer for access to the bottom of the fill area.

RIGHT-OF-WAY PIONEER

In advance of the right-of-way pioneer road construction, two sawyers with at least a week's head start must fell all trees larger than 6 inches in diameter along the road right-of-way (a term describing the construction limits of the roadway). The sawyers must drop the trees so they fall parallel to the roadway. When the pioneer dozer comes through it can easily turn the trees aside.

If the pioneer dozer attempts to push over the large trees, there is the hazard that one will fall back on the dozer. Breakage of trees pushed over by the dozer is common. This creates a problem because most contracts stipulate that all merchantable timber must be utilized. A contractor will be charged for any trees he breaks.

The right-of-way pioneer road might be construed as a violation of the contract because excavation is done before the clearing is completed. All contracts, however, stipulate that a tractor may build a pioneer road just large enough to give the clearing machines access to the project.

A problem often arises over the question: What does "just large enough" mean? The engineer, concerned about too much slash (stumps, tree tops, brush, etc.) being covered with dirt, wants the pioneer road to be barely wide enough to accommodate the smallest dozer. The contractor, however, realizes that the pioneer road must be wide enough for large clearing cranes and slash-hauling trucks. The end result is a compromise—a road wide enough for the larger construction equipment, but not so large as to cover all the slash.

A 12-foot wide pioneer road should suffice for most construction equipment. A small dozer should be used, but not one so small that it cannot root out the larger stumps. A dozer in the 170 horsepower class is adequate. Anything larger may be too big to operate on so narrow a road.

To begin, the pioneer cuts should be roughly gouged out of the hillside as shown in Fig. 3-1 since they will be obliterated once the earthwork phase begins. A contractor should make it clear to the dozer operator that slopes need not be smooth and uniform, the road itself should not be dwelled upon. Instead, the operator must gouge out a driveable platform and then move on. The best location for the right-of-way pioneer road is along the centerline of the proposed road. These centerline stakes will be in place because the road was originally layed out from them. This will result in an access road

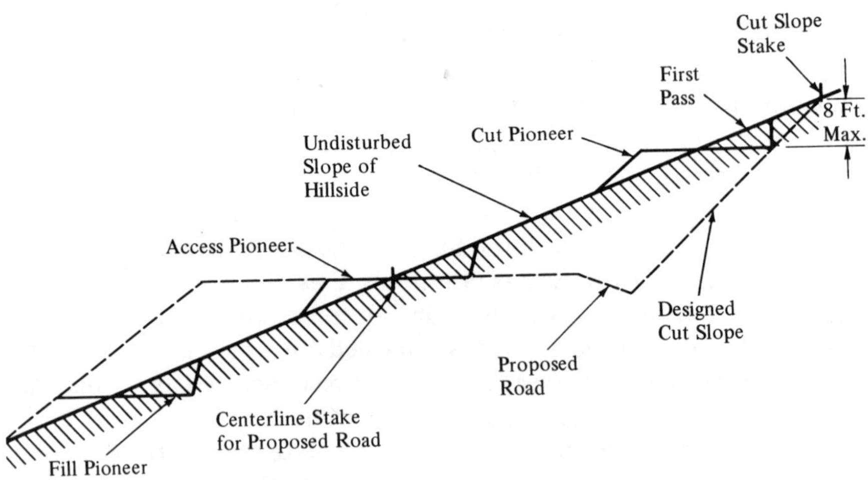

Fig. 3-1. Illustration of the three pioneer locations on a road project.

with roughly the same grades as the finished road. Contractors often try to save an extra operation by attempting to cut both the pioneer right-of-way and the slopes at the same time in order to use the cut pioneer as the access road. This is a mistake. The cut line on any road jumps up and down sharply depending on the terrain. The end result could be a pioneer road with many impassable grade pitches.

Streams and deep draws with running water pose a problem when constructing the pioneer road. Streams are protected waterways and therefore a contractor cannot drive his equipment through them. Instead, either a temporary or designed pipe must first be installed, then covered over with enough dirt to support the movement of construction equipment. An example would be a draw designed for 30 feet of fill at centerline with a 36-inch pipe at the bottom. In this case a contractor will save many dollars by going through the trouble of installing the designed culvert. This situation will require some clearing on either side of the draw so a dozer can push enough dirt into the draw to bed the pipe and provide adequate cover up to a level that permits a driveable grade. Installing a temporary culvert is more costly since the fill must be removed to get at the temporary pipe. This can involve one or two days of work for a hydraulic hoe.

When small creeks or draws are encountered with less than 10 feet of fill over them, it is sometimes easier to put in a temporary pipe and then fill over it a few feet. These temporary pipes can be old culverts pulled out from reconstruction of other roads. When the time comes to lay the designated pipe, a dozer or backhoe simply pushes off the fill and yanks out the old pipe.

Right-of-way pioneer roads are unnecessary for reconstruction projects since there exists an old road for access. This fact should be taken into consideration when bidding a reconstruction job.

CUT SLOPE PIONEER

After the clearing has been completed on a number of cut slopes, it is best to construct the cut pioneer road as soon as possible in advance of the earthwork operation. By doing this, a contractor can discover what types of materials are hidden inside the cut. Plans can then be made to bring in the correct size and type of equipment

Fig. 3-2. Large dozer begins pioneering a large cut slope on a recent construction job. Remember, no access pioneer is needed when reconstructing an old road.

such as ripping dozers, rock drills, etc. A large 240-horsepower dozer should be used to construct the cut pioneer road because rock or hard clay may be encountered and must be broken through to give the rock drills a drilling platform. Figure 3-2 shows a large dozer beginning a cut slope.

The cut pioneer dozer works a safe distance (about 8 feet) from the cut slope stakes for a length of about 100 feet at a time. This first pass is nothing more than a vertical gouge to give the dozers a working platform. The inside edge of the pioneer road should be about on the designed cut slope as shown in Fig. 3-1, upper right. The dozer then returns and slopes the vertical surface of the gouge back to the cut slope stakes of the designed slope. Since this is largely an "eyeball" operation, only the most experienced dozer operators

Fig. 3-3. This pioneer of a cut slope on steep, rocky ground caused undercutting of the slope stakes. Progress was further delayed because of rock.

do the cut pioneering. The dozer platform should not be lower than 8 feet below the cut stakes because that is the limit that a dozer can be expected to scrape back to the cut stakes. If the work cannot be done with a pioneer dozer, a backhoe must be brought in, a slow and costly procedure. Figure 3-3 shows a pioneer cut slope on steep ground that resulted in the undercutting of slope stakes.

FILL SLOPE PIONEER

Once clearing is completed in a fill area, a dozer of just about any size can construct the fill pioneer road. Some contractors don't even worry about it until the area is about to be filled. This is because the fill pioneer is simple for an experienced dozer operator to construct.

It is obvious from Fig. 3-1 (lower left) that the steeper the slope of the hillside, the more a dozer must cut into the earth to build a working platform. In so doing, the dirt is pushed aside and governs how close the pioneer road will get to the bottom of the fill.

A dozer on ground with a 10 to 20% slope can pioneer almost to the fill stakes, but on 40% ground the dozer may still be 10 feet above the fill bottom when the dirt from the pioneer road reaches the fill stakes. This creates occasional problems with the engineer because theoretically, every fill should be placed in layers and compacted. However, it is impossible to get a dozer to the bottom of the fill because it needs a level road from which to work. All contracts therefore, stipulate that fills shall begin where equipment can reasonably be expected to operate.

4 Clearing

The clearing operation begins after the access pioneer road has been built. There are several clearing methods that can be employed on a project. However, the nature of the terrain often dictates which method will be most efficient. Crane clearing is by far the most efficient operation on steep ground with deep draws to clear. When the slope is less than 25%, a crawler tractor equipped with a brush blade is unbeatable. For a project with some flat and steeper draws, the use of a brush blade tractor equipped with a winch and choker cables is efficient.

If possible, before the clearing operation begins in earnest, the merchantable lumber should be decked and arrangements made to haul the logs to the mill. Some unfortunate contractors end up hauling logs two or three times before they are removed. These logs take up space and hide an accumulation of tree tops and limbs. If the contractor clears an area and then must reclear after the log decks are removed, he is burdened with an unnecessary and expensive extra operation.

CRANE CLEARING METHOD

A 25- to 30-ton crane, equipped with a grapple and 60-foot boom, is a good rig for reaching down and hauling up slash from steep slopes. (See Figure 4-1.) At first thought, this capacity might seem excessive, but the many huge logs and stumps dug out by the pioneer

Fig. 4-1. A 20 to 30 ton crane is ideal for clearing on steep terrain. The large size is needed for picking up large stumps and logs.

dozer require a large machine to lift them. The job of the crane is to pick up the slash and stumps from below and above the pioneer road and deposit them on the road or in a truck. A crane is not designed for running along the road carrying a grapple full of wood, thus another machine should be used to transport the slash to the burn or disposal area. Many contractors utilize their existing equipment for this purpose. An end loader works well, as does a dump truck. However, a crawler tractor equipped with a tilting brush bucket and hydraulic clamps as shown in Figure 4-2 is the most efficient. A slick, muddy, outsloped road won't slow down this machine as compared to the rubber-tired rigs.

As the crane deposits the slash on the road, a sawyer should be available to cut up the longer logs and trees, making it easier for the

Fig. 4-2. A track-mounted machine with brush bucket and overhead clamps can haul a surprising amount of slash after the crane deposits it on the pioneer road.

hauling machines to handle them. The crane can't pick up every limb so it's a good idea to have a few "stick pickers" on hand. This is a term for laborers who stay a safe distance behind the crane and throw the remaining debris onto the shoulder of the road. When time allows, the hauling machine drops its bucket and runs along the shoulder scooping up the wood.

The one glaring inadequacy when clearing steep ground with a crane is that all in-place stumps must be left. In other words, clearing is not totally complete even though the area is ready for the earthwork operation. However, this practice of leaving behind the in-place stumps is standard procedure for efficient road jobs on steep ground.

The earthwork dozers root out the stumps with no extra effort as the cuts and fills are developed. The stumps are kicked down the hill past the limits of the fill area. When the road is finished, the crane makes a quick pass through the project and loads these stumps into a truck to be transported to a burning area.

A stump cruncher can be extremely useful because of the difficulty in burning stumps. A stump cruncher on a crane works similarly to tree pruning snips. The stump cruncher picks up the stumps and deposits them on the pioneer road and then splits them in half. This makes a great difference in burning efficiency and these split stumps disintegrate along with the other lighter slash. Another variation of the stump cruncher can be installed behind a clearing dozer. A stump is deposited on the pioneer road and the dozer then backs into it, pushing it up against the cut slope and splitting it. However, the stump cruncher on a dozer is not as quick or efficient as one on a crane. In some clearing operations a crane goes through and picks up only the slash and leaves the stumps behind. As it leaves, another crane comes up and crunches stumps. This is an expensive procedure and reflects the concern contractors show toward disposal of stumps.

Crane clearing on steep ground gives rise to the problem of disposing of the slash to a suitable place for burning. When the slope is constantly steep, burn pits are necessary. These are excavated pits dug out by a dozer in which the clearing trucks or loader dump their slash. These pits are located outside the lower roadway limits and spaced anywhere from 2000 to 4000 feet apart, as designated by the engineer. It is impossible to efficiently clear steep ground with a crane without the use of these burn pits, and because they are outside the immediate fill area, it is not necessary to work around them.

From the description of this operation it appears that the crane would quickly outperform the hauling rigs, but a tractor with brush bucket can keep up with a crane on a 1500 foot haul. It sounds like a slow, laborious way to clear, but when it is set up right, the crane clearing method is far superior for a tight, steep job.

BRUSH BLADE METHOD

Besides speed, another feature of the brush blade method is the low investment cost. Any crawler tractor can have its dozer blade dropped and a brush blade installed. This is shown in Figure 4-3. Thus a contractor can clear and build a road with the same machine! There are, however, some limiting factors to this type of machine. A brush blade tractor simply cannot push clearing slash very far be-

Fig. 4-3. On flat ground a dozer with brush blade is most efficient. Here the dozer roots out stumps and pushes them into a pile to be burned.

cause stumps and logs continually work loose from the blade. The ideal situation is to have a burn pile or disposal area about every 500 feet. That means a maximum 250-foot push each way. Even then, the brush blade tractor will have to make several passes after lost wood.

Sometimes, because of burning regulations or designated burn areas, the distance between slash piles may be increased greatly, making the clearing job uneconomical unless the brush blade tractor gets some help. Here again, a rubber-tired end loader or tractor with brush bucket can be used. The brush blade machine simply roots up all the stumps and slash and piles them in the middle of the right-of-way every couple hundred feet to be carried off by a hauling machine.

Aside from its obvious simplicity of operation and speed, the brush blade method leaves the right-of-way completely cleared. The converted dozer has the power to root out the stumps so the earthwork operation can follow with no difficulty. Another advantage is that there is no need to come back to pick up stumps that were kicked out during the excavation phase. A laborer or two with saws can complete the operation.

BRUSH BLADE/CHOKER METHOD

Often a jobsite will have gently sloping ground on the sidehills, but the draws will be much steeper. The brush blade tractor cannot push large loads up steep slopes and may even have trouble itself getting up these slopes.

A smart contractor just doesn't fight it. The brush blade tractor operator backs down to the clearing slash as close as he dares, and then reels out a winch line to a choker setter, a term given a laborer who sets noose-type cables. The choker setter usually has four or five chokers wrapped around stumps or logs which he then hooks to the winch line. The brush blade tractor skids the slash to the burn or disposal area, where another laborer is waiting to unhook the chokers. While the tractor is skidding the slash the choke setter is wrapping more chokers. After unhooking chokers in the burn area they are carried back to the clearing area by the returning tractor. The brush blade tractor is never sitting idle waiting for someone to set or unset chokers. It sounds and often looks like a ponderous process, but leave the area for awhile and return—it's amazing how much wood has disappeared.

BURNING

Burning would appear to be the easiest part of the clearing operation. Take, for example, a large pile of wood and stumps, say 20 feet in diameter and 10 feet high. The pile is ignited and it burns fiercely for several hours before dying down. What's left? Stumps! Stumps create a problem throughout a job. They just won't burn easily. It takes a huge pile of slash to produce the heat intensity required to disintegrate stumps. Contractors will often haul slash longer distances at greater cost and build several huge burn piles instead of many small ones.

Diesel fuel once was the standard for starting a fire in a burn pile. When the fuel was cheap, gallon after gallon was thrown on a kindling fire to ignite it. But with the great increase in fuel prices, contractors have begun raiding junkyards for old tires. One or two truck tires and a gallon or so of diesel fuel are enough to start even a wet burn pile.

Sometimes in a crane operation where the clearing is light and the crane is continually moving along the right-of-way, a metal burning box is used. The box is usually homemade, about 10 by 20 feet by 8 feet high, constructed of ⅜-inch thick welded steel plate. Such a box pulled along by a small dozer is shown in Figure 4-4. Intense heat is produced from the compact mass confined in the box, and stumps readily disintegrate. Burn boxes can be equipped with a device that continually sprays a fan-forced stream of diesel fuel into the bottom of the box through the lower air holes. Burning is intense and total, but the cost of operation is prohibitively high unless it is called for

Fig. 4-4. A steel plate burning box is used when clearing is light. This box can be periodically pulled along by a small dozer.

in the contract, whereupon the contractor would be compensated for such extraordinary measures.

A clearing crew progresses more slowly than an earthwork crew, so it is a good idea to have the clearing crew at work for about a month before the earthmoving operation begins. Many contractors feel that clearing is a necessary evil in order to get at the more lucrative earthwork phase. But with a little planning, clearing should be profitable if the system that is selected uses manpower and equipment efficiently and takes into consideration special features of the terrain.

5 Roadway Excavation

Roadway excavation is the real money maker on a construction project. It usually pays the highest, and contractors readily accept losses from other phases of construction with the intent of making them up during the earthwork operation. Sometimes they succeed, other times not. As mentioned previously, every item of work should be bid and executed with the intent to make a profit. If a contractor can make a few dollars from these other operations, he will be all the more ahead when it comes time to move the dirt.

Contractors new to road building hear discussions about the lucrative earthwork operation, and expect money to automatically fill their pockets. This is just not true. Knowledgable contractors profit from the earthwork operation only because they have perfected it to a science. They may simultaneously run several earthwork operations during the course of one project. Trucks, dozers, loaders, hydraulic hoes, scrapers, and other equipment are utilized fully to increase efficiency and thereby increase profit.

To understand the efficient use of earthwork, it is imperative to understand how excavation is designed. A road project is broken down into many small sub-projects. Within the boundaries of each of these sub-project areas, the excavation is designed to build the fill, that is the material removed is used to fill in low spots in the roadway. On either side of the area is a balance point. Between balance points, which is a measured distance along the route of the road, the excavated material is equal to the needed fill. On the design drawings, balance points are marked as "B.P." and the quantities of cut and fill are listed for each area (Fig. 5-1).

A word of explanation is in order for the design drawing. It shows the superimposed profiles of the finished slope and starting slope. The area above the finished slope line represents the material to be removed, the area below is for fill. The numbers to the left (3950 to 3990) are elevations. An 18 inch diameter culvert is shown at the upper right.

Below the slope profiles are the Balance Point locations. The first B.P. is at station 1 + 06, the second at 5 + 41. The numbers above and below the arrows show the excavation and fill amounts, respectively. In the first case 156 cubic yards of excavated material will yield 139 cubic yards of fill. The difference is due to shrinkage of earth.

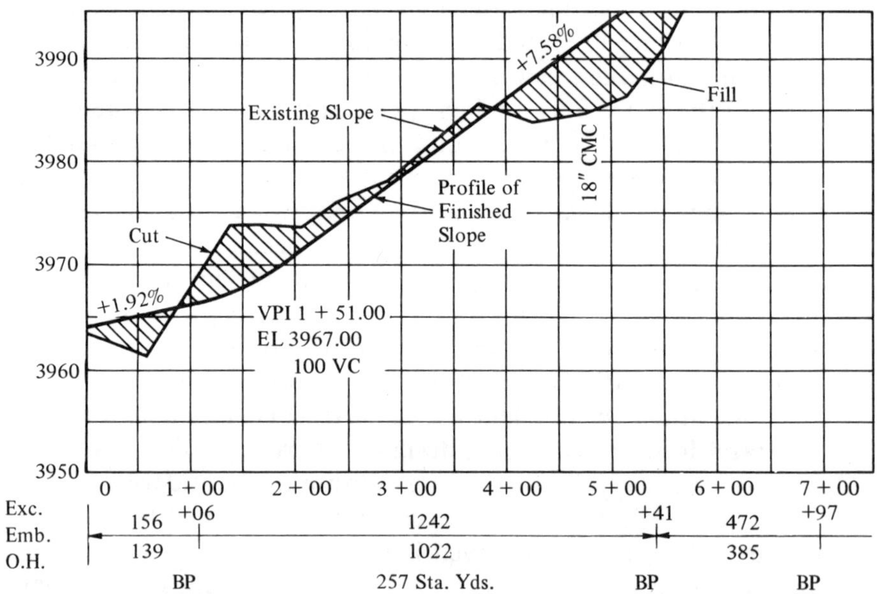

Fig. 5-1. A road project is broken down into many small projects in which the cut within a certain area will build the fill in that area. These are called Balance Points. The first balance point in this illustration is at station 1+06; 156 cu yd of excavation builds 139 cu yd of fill.

There are some common abbreviations found on design drawings.

EXC. : excavation
EMB. : embankment
O.H. : overhaul
V.P.I. : vertical point of intersection
EL. : elevation
V.C. : vertical curve

A vertical curve is used to ease the transition between two different grades. The point where two different grades meet is called the *vertical point of intersection.*

These balance points are what a contractor notes when bidding a job because the greater the distance between balance points, the more expensive it will be to move the dirt. For example, a 300 foot balance point is ideal because a dozer can push the dirt that distance economically. A balance point of 1000 feet, however, may require the use of a loader, trucks, and dozers. Obviously this operation is more expensive. A balance point that is 3000 feet long is beyond the economical range of trucks, so scrapers with their quick turn-around time are used, but only if the excavated quantity is large enough to warrant such expensive earthmoving equipment.

A small contractor with one dozer who expects to make big money on a large construction job is in for a surprise. Big jobs are designed for large contractors, and much of the work is purposely designed to take advantage of a fleet of trucks and scrapers.

You might ask if the small sized or beginning contractor has any business in the road construction business? The answer is a resounding YES! There are always 30,000 or 40,000 yard road jobs that are ideal for small contractors with only a dozer or two. These roads are designed with the idea that probably all excavation will be moved by dozer. Many construction companies started out with one dozer and worked up to a multi-unit operation.

CLEARING

The contract stipulates that clearing must be completed before earthwork operations can begin. This often misleads the contractor to believe that clearing must be complete for the entire project, or at least a sizeable portion, before excavation can begin. This is not true. Clearing must be complete only between the balance points for the excavation work to begin between these points. A contractor may construct the road right behind the clearing crew if he so elects. In fact, this is what usually happens toward the end of the project when the excavation operation catches up with the clearing operation.

SLIVER CUTS AND FILLS

There are cuts and fills that are impossible to build because the ground is made up of thin slivers of material that no dozer can work. On sliver fills, a dozer may be able to pioneer down only a few feet before the loose dirt falls down to the fill stakes. On a 20-foot sliver fill, the dozer may still be 15 feet above the fill stakes. An experienced engineer is helpful in such situations. He can recognize a sliver cut and fill and waive the specification. An inexperienced engineer may require the contractor to make one or more attempts to construct in the area in question, but still, a contractor is bound by contract to construct only what is practical to build.

At times sliver cuts can be scraped down with a large hydraulic hoe. But in rock, a drill may be needed to put in some side holes for blasting. The use of explosives can leave a ragged cut line with overhang at the top of the cut. This is not the contractor's responsibility but rather the fault of the design. Nothing should be designed into a road that cannot be "practically" constructed.

COMPACTION

If no compaction is called for in the contract, the fill is placed in whatever wet or dry condition it is in at the time of construction. However when a compaction requirement appears in the spec, as it does in highway jobs, the contractor must compact the dirt to a certain percentage of its maximum density, usually 95%. This sounds high and extremely difficult to obtain, but it is not. If the soil is moist, a pass over it with a roller should be adequate to get 95% compaction.

Water is the key to any compaction effort. If the soil is dry, no amount of rolling will compact it. When low moisture conditions appear during excavation, it is advisable to have a water truck constantly watering the work area. The watering can be done at the cut or fill areas. Most often, however, the fill area is watered because the cut is often too steep for the water truck. Also, when the material is dumped into the fill, it is loose and more easily soaks up the water. Figure 5-2 shows a water truck in operation. Figure 5-3 illustrates how a dozer is used to pull a grid roller for compaction. A closeup of a grid roller is shown in Figure 5-4.

Fig. 5-2. Water is the key to any compaction effort. When soil is dry, no amount of rolling will compact it.

Fig. 5-3. Compaction makes the earthwork operation more complex and expensive. Here a dozer must be used to pull a grid roller for compaction.

Fig. 5-4. The grid roller is used to compact rocky fills because of its many pressure points which pulverize large rocks.

SLOPE AND GRADE STAKING

The contract requires that all cut and fill slopes be constructed to their designed slope as written on the construction stakes. For example, a cut slope noted on its stake to be excavated on a 1:1 slope means that for every foot down, the slope goes out 1 foot. On a 2:1 cut slope (used in shallow cuts), for every foot down in elevation the slope moves out horizontally 2 feet. In other words, a 2:1 slope is half as steep as a 1:1 slope. This subject is covered in more detail in Chapter 15.

If cut slopes are excavated at a steeper angle than designed, it means that hundreds or even thousands of yards of material are being moved for which the contractor will not be paid. Since this steep angle is a violation of the contract, the engineer may require that dirt be laid up against the steepened slope to return it to the designed angle. Obviously, this is an expensive mistake to make.

Even on smaller road jobs, a stake setter, or stake hop, is employed to do nothing else but continually check cut and fill slopes. He need not talk to the dozer operators at all. He simply sticks a lath into the cut slope every 5 to 8 feet down the slope and the dozer operator can see for himself if the slope is too steep or too flat. Seldom will a slope be constructed exactly as designed. Instead, slopes continually vary a few tenths of a foot. This is normal because a dozer operator must first be concerned with production and second with accurate slopes. That is why the stake setter is so valuable. He keeps the dozer operator constantly informed without slowing down production. A stake setter can keep several earthwork operations going simultaneously since the procedure for checking slopes is quick and simple. On smaller jobs, the stake setter often has other duties such as the driver of the water truck or foreman.

DOZER EXCAVATION

Excavation and embankment are done by a dozer when the balance points are close enough to make it economical. A dozer in the 270-horsepower range equipped with hydraulic ripper teeth is ideal. The dozer blade should be slightly U-shaped (commercially called a *U-blade*). The U-blade on a large dozer can push about 7 yards because the blade is designed to keep the dirt from falling out the sides. A large dozer in good mechanical condition can move about a thousand yards a day for short hauls.

Sidecast Earthwork Method

In this earthwork procedure, the dozer excavates the material and simply pushes it over the side until the fill slope is pushed out to the fill stakes. Slope checking on the fill side is unnecessary on sidecast fills. Neither is a fill pioneer necessary, and stumps in the fill area can be filled over without uprooting them. Any stumps rooted out during cut excavation can be pushed out of the work area and off to the side until they are picked up by the crane after the excavation is completed. The sidecast method is the simplest of all road construction, but the pay is less than more difficult excavation methods.

Layer Placement

Layer placement of excavated material into fill areas is more complex than for the sidecast method, thus the pay is higher. In this method the dozer pushes the dirt out of the cut area into the pioneered fill area in layers (called *lifts*) of about 1 foot, depending on the specs. When the area to be filled is just below the cut, the dozer must sidecast about 50 yards over the side onto the fill pioneer and then go down to the pioneer and layer place it.

When compaction is specified in the contracts, the earthwork operation becomes more complicated and expensive because the fill must be compacted between the 1-foot lifts. This is usually done using a heavy roller, either a steel drum type or sheeps foot or with a grid roller. The roller is pulled by a smaller dozer as it is not advisable to slow up the excavation dozer. Also, the narrow confines within which the larger dozer works makes pulling a heavy roller unsafe.

With two dozers the earthwork operation proceeds much more rapidly. The large excavating dozer has only to push the material to the fill area before returning for another load. While it is gone, the fill dozer with the roller places a 1-foot lift into the fill area and compacts it.

When a dozer pushes dirt for long distances, the material tends to dribble out both sides of the blade. The dozer should purposely let the material dribble out for several pushes, and then push a trench through it, leaving a 3-foot shelf or berm as it is called. When the dozer pushes a load of dirt, it cannot dribble out because the high berm holds it in place. This is a critical technique when pushing material more than a hundred feet. A berm is shown in Figure 5-5.

When rock is used to construct a fill, the depth of the layers to be placed changes to 2-foot lifts. The contract will stipulate this depth, the reason being that rock needs very little compaction and will not settle. Rock-fill slopes should be more closely inspected by the slope checker because rock tends to stand steeper than dirt. If rippable rock is encountered in the cut only a few feet above the finished road grade, it is advisable for the dozer to rip it in order to finish that section and continue on. When rippable rock is encountered 10 or 15 feet above road grade, production will be cut drastically if the dozer must make two or three passes to rip the rock. When the

Fig. 5-5. When a dozer must push material more than 100 ft, a berm should be built up, and the dozer can then push efficiently in the trench.

Fig. 5-6. A ripper is a 4-in. thick pointed steel tooth which is used to loosen up soft or fractured rock.

quantity of rippable rock reaches 100 cubic yards or more, it is best to drill and blast it. Remember that ripping rock puts an extremely heavy load on a dozer, and causes immediate maintenance problems. Figure 5-6 shows a ripper attachment for loosening soft or fractured rock.

A few drill holes should be enough to break up rippable rock but the drilling time is slow. There is no set rule to help decide whether an area should be ripped or blasted. A short cost analysis in a few areas will help the contractor determine the most economical method.

EXCAVATION BY TRUCK

When balance points extend several hundred feet or more, the efficiency of the dozer rapidly drops off. If there are only a few of these balance points, a scraper cannot compete with the operating costs of a truck-haul setup. However, don't be misled. A truck-haul operation is not cheap because of the additional construction equipment required.

The cut area to be truck-hauled must still be manned by a dozer which excavates the material and feeds it to a front end loader which in turn loads the trucks. See Figure 5-7. The push dozer is mandatory because an end loader is not made for excavation, and can operate only when fed loose material already excavated by the dozer. Another dozer must be on hand in the area where the trucks are dumping in order to layer-place the material. Figure 5-8 shows the truck dumping area and the dump pile that must be spread by the dozer.

When working in common dirt or gravel a loader in the 170 horsepower range should be used but when larger rock is encountered, a 325 horsepower loader is advisable. What is the best size truck to use? The answer is any size available. A standard small dump truck has a 10-yard capacity, but the average hauled load is only 8 yards. Larger trucks with a 16-yard capacity are better, and the 20-yard

Fig. 5-7. Hauling excavation by truck is expensive and is used when overhaul is paid to compensate for the extra equipment.

Fig. 5-8. A dozer must be on hand at the area where the trucks are dumping to spread out the dump piles.

rock trucks with steel dump beds are the best. The large contractor has the option to use a variety of truck sizes, but the small contractor must use what he has. Any truck can be used to haul excavated material.

When small trucks are used to haul rock, their dump beds and hydraulic hoists are easily damaged from the constant pounding when the loader drops the rocks several feet into the bed. This can be prevented if the loader will scoop up a half load of fine dirt and drop it into the truck bed. The dirt will cushion the blow when the large rocks are dropped into the truck bed.

Trucks need a safe, level platform from which to dump. This platform is formed by the fill area dozer. Often, an area to be filled has only a 10-foot pioneer road through it, barely wide enough for a truck to drive on, let alone end-dump over the side. In the beginning of the operation the trucks dump their loads on the pioneer road and the fill dozer side casts the fill until a safe platform is built from which to dump. It is advisable to have a man at the dump site to direct the trucks to the edge of the platform and to keep a critical eye on the loads that are being hauled. A truck with most of the load on one side may tip over when the bed is raised. In a truck-haul operation, every piece of equipment should have something to do at all times. The greatest inefficiency results when there are not enough trucks to keep the other equipment working. Two trucks may be fine for a 500-foot haul, but when the haul is a distance of a thousand feet, four trucks are best.

The use of two-way radios is imperative for haul trucks. By keeping track of each other the trucks have fewer accidents. Also, the truckers can decide who will pull off the road to let the other one pass, thereby increasing production. The standard rule is that the loaded truck always has the right of way on the more stable inside track.

The truck haul method is far more expensive than the dozer method. Just analyze the equipment. In the dozer method all that is required is one or two dozers, but for the truck-haul method at least two dozers, an end loader, and a number of trucks are required.

Contractors, however, do not shy away from road projects just because haul is involved. The reason is that they get paid for it, sometimes lucratively. For example, a road designed for a 300 foot freehaul distance has no extra compensation for the reason that a dozer can accomplish such haul efficiently. After 300 feet, the contractor is compensated for all excavation. It is not unusual in long hauls to have the haul cost approach the excavation cost. In other words, it may cost $1.00 for a dozer to excavate a cubic yard of material and another $1.00 to haul it.

EXCAVATION BY SCRAPER

For a 4000 foot distance, two scrapers can haul more excavated material than 6 dump trucks. This is an amazing statistic. Scrapers can roar along at 40 miles an hour with a full load, dive right down into a fill area, lay down a 1-foot lift, and head back for another load. Figure 5-9 shows a 4500 cubic yard cut excavated in only 7 hours by two scrapers. They are the elite of the earthmoving corps. Why then are they not used for all earthmoving situations? The reason is that they are expensive to buy and maintain.

Fig. 5-9. Two scrapers excavated this 4500 cu yd cut in seven hours. Nothing moves dirt as fast as scrapers.

Scrapers range in size from 10-yard single units to 40-yard double-engine machines. However, the 20-yard scraper is most generally used. The 40-yard scrapers are difficult to turn around and maintenance costs are high. Scrapers can excavate common dirt, but in hard packed gravel, clay, or rock they require help and a large dozer must be on hand. The scraper pulls into the excavation area and drops its blade; the dozer places its blade against the rear of the scraper and pushes it ahead until the material piles up in the container of the scraper. Twin-engined scrapers can excavate much more than single-engined units without the help of a dozer, but when rock is encountered, even they need help.

When shot or loose rock is encountered, the operator of the scraper must be careful not to pick up any large rocks. A 2-foot diameter rock can be forced into the hull of the scraper, but will not disgorge as easily. This often holds up the entire operation while a team of men with pry bars wrestle large rocks out of the hull. The push dozer operator should watch for large rocks and push them aside. If the entire excavation consists of large rocks, a fleet of trucks may be needed.

A fill dozer with roller is not required with scrapers because the huge scraper tires exert tremendous pressure on the fill material. If each scraper splits the tracks of the preceding machine, compaction of the fill will be adequate. This is a big money saver often overlooked.

Since scrapers are expensive, all long haul areas should be ready and waiting for their arrival. This may mean leaving a few areas undone when the regular earthwork operation goes through. It is sheer economic disaster to bring in a fleet of scrapers two or three times on one project.

EQUIPMENT RENTAL

A contractor with only a few pieces of equipment is often frustrated when looking at a prospective job which has mostly dozer excavation, but with a few areas having 1000 feet or longer hauls. Without a loader, trucks, or scrapers, it would be impossible to move the dirt. Smart contractors still bid these jobs, do the bulk of the work with their own equipment, and rent the equipment needed to do the special work.

There are several sources of rental equipment. Every locality has a firm which rents heavy equipment by the month. A month's notice is usually necessary since this equipment is constantly being rented out. This is the touchy area. If a contractor tells the rental firm that he needs two scrapers for the month of August, the areas must be cleared and ready for them. He cannot readily change dates with the rental firm if his schedule has slipped.

Renting equipment by the month often loses money for a contractor, but this loss is usually expected. For example: two scrapers are rented for one month at a total cost of $20,000. The excavation and haul that they are to do amounts to only $16,000 in pay and takes only two weeks to complete. But if the rest of the job has large cuts which are dozer-excavated, the contractor will purposely take a loss of $4,000 on the scrapers and make up for it with his dozers.

Equipment can also be rented from other contractors. Scrapers and trucks, for example, can sit unused for months in a large contractor's storage area. A call to a few local contractors is bound to produce the equipment needed, and rental can be made by the week, which is a big money saver.

Trucks and loaders can be rented from a gravel plant or small residential building contractor, but keep an eye on the operators of this equipment. They are not used to the preciseness of major road construction and are apt to sit chatting while the support equipment waits for them. Scraper operators, on the other hand, normally are no problem. On the contrary, if there is any problem, it's in slowing them down.

Equipment rental is an invaluable aid to the contractor. Even large firms rent equipment such as hydraulic backhoes. Some contractors who bid large construction projects own only a few pieces of equipment, choosing instead to rent most of it. Equipment such as scrapers, hydraulic backhoes, cranes, gradealls, and rock drills are so specialized that most contractors cannot keep them working even half the time. Consequently, such machines are most often rented.

6
Culvert Installation

Many contractors doubt that it is possible to install culverts at a profit. While it is true that most contracts do not include lucrative culvert prices, a contractor can still make a profit if culvert pipes are installed in a systematic, efficient manner. However, no profit will be realized if a culvert crew installs only one, 30-foot ditch relief culvert per day. In order to make money at least 80 feet of pipe must go into the ground each day. This is not out of reach of an efficient culvert crew.

Certain culverts are truly necessary evils on a project, especially small diameter pipes that are laid on the bottoms of steep draws, or in creeks. A culvert crew may labor all day and well into the evening to install one of these, and still need the help of a dozer or two, a crane, and a hydraulic backhoe. It is impossible to make money on such installations, but a contractor can at least cut the loss to a minimum by using the most practical and proven construction techniques and make up for this on the other types of culverts required for the job.

DITCH RELIEF CULVERTS

Ditch relief culverts, or cross-drains, are small diameter pipes (18 to 24 inches) used to drain water from a roadside ditch underneath and away from the road. Various types of culvert installations normally encountered on a project are shown in Figure 6-1. Most cross-drain

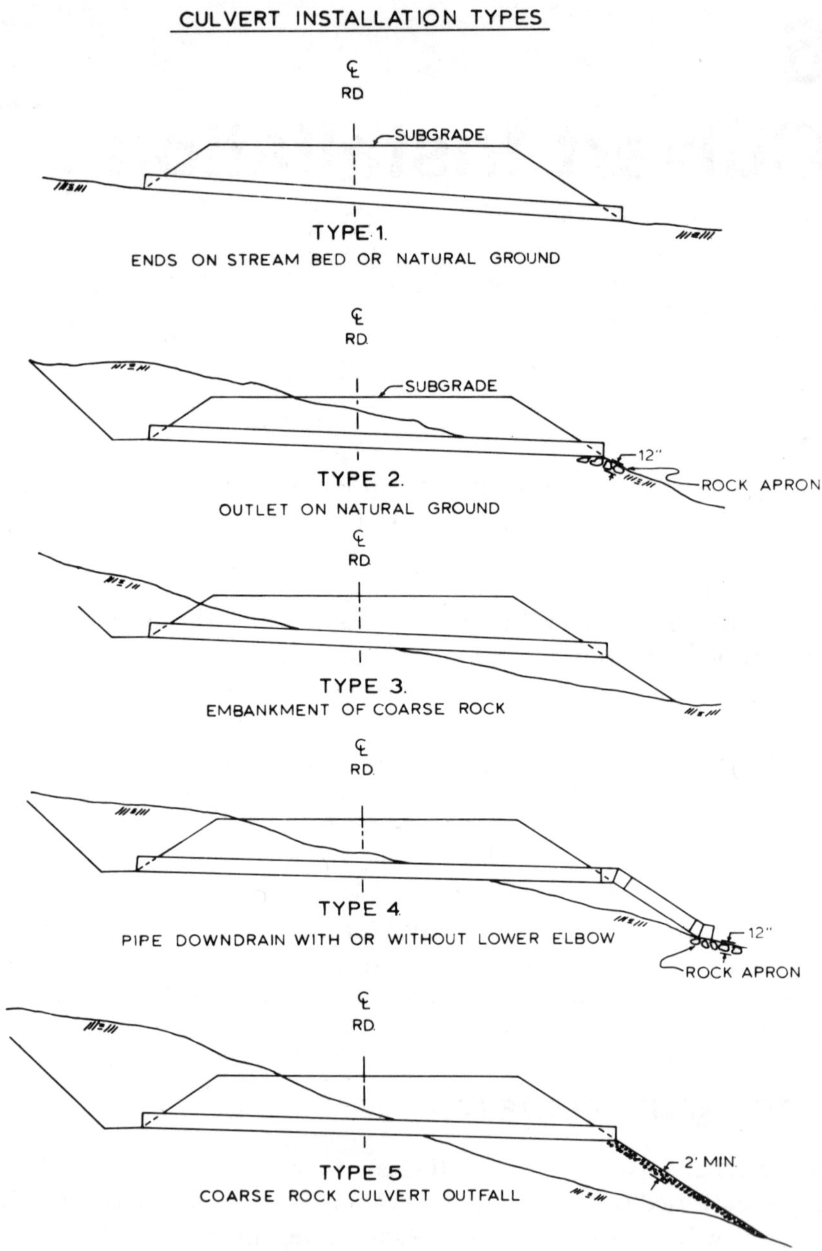

Fig. 6-1. Types of culvert installations normally encountered on a project. Note the water catch basin at the inlet on cut sections.

installations are simple, and the only time a problem arises is when rock is encountered when excavating the culvert trench.

Before the culvert crew begins work, the foreman should get together with the engineer and his inspectors and review the culvert specifications and the installation procedure so that when the actual work begins there will be no confusion as to the correct procedure. The foreman can then inform his men. This is the best way to get the culvert work underway without having the first few rejected because someone didn't understand what was required.

A common mistake with a new culvert crew is to install a pipe with the lip of the pipe facing into the flow. This presents a ledge for the buildup of debris and can result in blockage of the culvert. Figure 6-2 shows an enlarged portion of a culvert joint and the proper position of the inner lip with respect to flow.

A typical culvert crew should be comprised of one backhoe operator, two laborers, and a foreman (the foreman also works as a laborer). Three laborers are desirable because the strenuous nature of this work exhausts two men before the day is half over, and production decreases. Each man should be equipped with a shovel, and there should be a few spare shovels available since the handles break

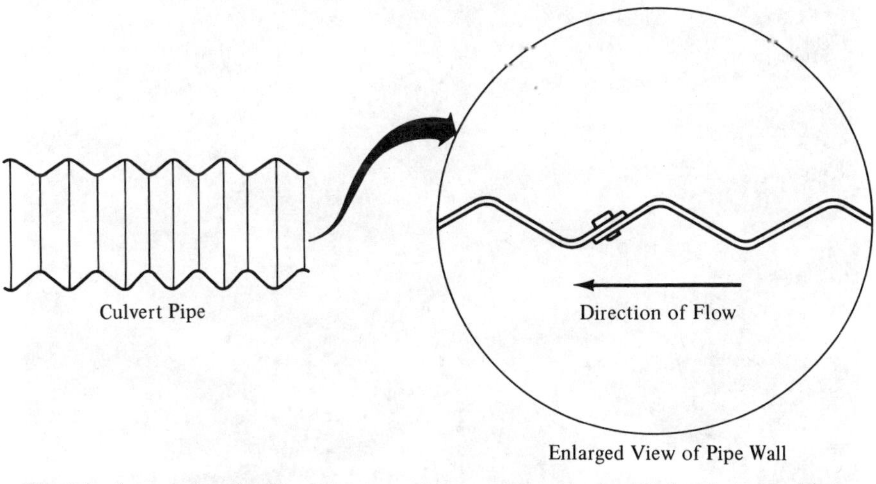

Fig. 6-2. A common mistake with a new culvert crew is a pipe installed with the outside lap pointing downstream instead of upstream. If the inside edge is pointing upstream, debris tends to get caught on the lip and blockage of the pipe is possible.

easily. A couple of ratchet wrenches are also needed to bolt the connecting bands tightly around culvert sections that butt together. A small flatbed truck or three-quarter ton pick-up is used to transport the culvert crew and its equipment, and to haul culvert pipe from the main stockpile to the installation sites. A small size backhoe in the 70 horsepower range should be used for excavating the pipe trench. It should be equipped with a one-quarter cubic yard backhoe bucket, and a 2 cubic yard front end bucket.

In order to install a pipe at the designed grade and depth, a hand level is necessary. This is a hand-held eye level that gives a horizontal line-of-sight when a small bubble is centered in a water-filled tube. Elevation differences are measured by sighting through the hand level from a reference point to a grade rod. These are in constant use to find the inlet, grade, and outlet elevations of pipes. Remember, in construction work there is no such thing as an inch; all measurements are in tenths of a foot.

The installation of a culvert begins when the backhoe excavates the trench for the pipe. Such a backhoe is shown in Figure 6-3. This

Fig. 6-3. A small backhoe is ideal for culvert installation.

trench should be a little deeper than the pipe grade because 4 to 6 inches of bedding material is required under the pipe. A worker using the hand level and rod can save a lot of time and guesswork by continually checking the trench elevation as the backhoe works. While this is being done, other workers should be loading the correct length of pipe onto a truck at the main culvert stockpile. When the trench is roughly to grade the laborers jump in to smooth out the humps and dips while the hand level and rod are used to check the final pipe grade.

Culvert installation usually calls for bedding material. The specs may stipulate that the bedding material can be dirt with no rocks in it larger than 2 inches, or bedding sand may be required. If sand is used, it should be dumped at each culvert site prior to the arrival of the crew. This bedding material is used to protect the pipe from any unseen large rocks that may push up through the metal. When the bedding material is spread out, the culvert pipe is laid upon it and the final grade checked again before backfilling begins. Figure 6-4 shows laborers removing rocks from a culvert bed as a backhoe

Fig. 6-4. Culvert installation in rocky soil is difficult. Here the laborers remove rocks from the culvert bed and the backhoe dumps in fine bedding material.

dumps fine bedding material. Most culverts are made in 20 to 30 foot lengths, so a connecting band must be bolted to the butt ends of the pipes if a longer length is required.

The pipe is now ready to be backfilled in the specified lifts. The backhoe deposits the fill material along both sides of the pipes in one-quarter yard piles, and the laborers smooth out the fill and pack it under the haunches of the culvert. A gas-powered hand tamper should be used to compact the fill. One laborer holds and directs the tamper, and another pulls it along using an 8-foot long rope attached to the bottom of the tamper. Figure 6-5 shows the compacting of soil around and over the culvert; the bed of the culvert is shown in Figure 6-4. Incorrect compacting of this fill material has the tendency of bowing the pipe along its length to the point where it is impossible to see through it. This can be avoided by tamping each lift on both sides of the pipe. Never tamp only one side all the

Fig. 6-5. The same pipe as in Fig. 6-4 is compacted. Note the lack of rock in the tamped bedding material.

way up. One of the laborers should periodically check the culvert for any signs of shifting.

If the bedding material is too dry, it may be necessary to water it in order to obtain the required compaction. A water truck can be used to water down the fill material. However, too much water should not be used because dirt can become mud quickly when a hundred gallons of water is dumped on it.

After the pipe is backfilled to the required depth, the backhoe, using its front bucket, proceeds to build the road over it, while the crew goes for more pipe. By the time the men get back, the backhoe should be finished excavating the next culvert trench. Culvert crews take a few days to synchronize their activities, but once synchronized there should be no trouble to install three ditch relief culverts each day.

Fig. 6-6. When rock is encountered during culvert excavation, the crew will be held up for hours. It is best to have some make-work ready for these situations. It took six hrs. to drill and blast this rock.

Rock is the one thing that can delay a crew. If hard rock is encountered in the trench, a jackhammer or airtrack drill must be brought in to drill and blast the rock. While this is going on, the culvert crew can install headwalls or rock outfalls on previously installed culverts. A foreman should always have some "make work" available for his crew when these delays occur. Figure 6-6 shows how rock can cause problems. It took 6 hours to drill and blast the rock in this trench.

CULVERTS IN DRAWS AND LIVE STREAMS

Culvert installation in draws and streams is a slow, frustrating job that rarely shows a profit. The best approach for a contractor is to get these installations over as quickly and efficiently as possible. Unlike the ditch relief culverts, which always follow the same installation procedure, placing pipes in draws and streams gives rise to such peculiar problems as underground springs, inaccessibility, muddy beds, etc. A contractor must analyze each problem and adapt his installations to fit.

In dry draws that lie on about a 30% slope or less, culvert installation is not too difficult, unless a large pipe is required. A 48-inch diameter steel corrugated culvert is impossible for four men to move even in 20-foot sections, and a crane must be available for no other purpose than to feed the pipe sections to the crew. Smaller pipes can be handled by the men.

Steep draws are real problem areas. Depressions along the pipe grade must be filled and humps cut down, tasks too great for men with shovels. In such instances a large hydraulic hoe can be worth its weight in gold as some of the larger models can reach almost 40 feet. Once the irregularities in the pipe grade are roughed out, laborers can finish it, and the rest of the installation is about the same as ditch relief culverts.

Occasionally, a steep draw is encountered that is out of reach of both the crane and hydraulic hoe. A large dozer can go down the draw, but cannot make it back up. In this situation, a small dozer (a landscape dozer) can be used to shape the pipe grade as it works its way down the draw. Once at the end of the pipe bed, a large dozer winches the smaller one back to the upper end, and starts all over again until the pipe grade is roughed out.

Culvert installation in streams presents special problems. The contract will call for the culvert bed to be a dry installation even if it is in the course of a stream, and the contractor must somehow divert the water temporarily out of the working area. If the stream bed is not too narrow, a dike can be built and the water diverted to a ditch alongside the pipe. But if the stream channel is narrow, a contractor is better off to dam up the stream and pump the water around the work area. This is the simplest way to get rid of water, and a 3 horsepower pump will handle a surprisingly large volume of water.

There are times, however, when a narrow stream channel gushes such a torrent of water that even a number of pumps cannot handle the flow, and all practical efforts to divert the stream fail. The contractor and engineer must both recognize the seriousness of the problem and work together to obtain the best installation possible. It is a fact that culverts have been installed in wet beds, and while the specification calls for a dry installation, the contract also specifies that all areas of work must be possible to execute in a practical manner. In other words, if the installation is that complicated, the contractor may be working outside the scope of the contract, and further compensation can be demanded.

ARCH PIPES

An arch pipe, or squash pipe, is an elliptically shaped culvert made especially for streams with large volumes of water. The squashed shape allows a greater area for increased volumes of water to pass, and has the added advantage of allowing the subgrade of the road above it to be lowered in critical areas.

Arch pipes are installed in the same manner as round pipes with the exception that the "haunches" of the arch pipes are extremely critical areas and should be tamped by hand with 2×4's if necessary to keep the pipe from squashing and structurally weakening it. The haunches are the undersides of pipe that do not rest on the ditch bottom and are unsupported until earth is filled in and tamped.

MULTI-PLATE ARCHES

A multi-plate arch is an open-bottomed culvert delivered to the project in pieces. These structures come with all the necessary steel framing, multi-plates, and nuts and bolts for assembly in the field. Multi-plates are large size culverts, 6 to 20 feet in diameter, and are made of a heavier gauge metal than standard culverts. Do not attempt to assemble a multi-plate arch without consulting the instructions accompanying each arch. Because of the elliptical rather than round-shaped arch, it is very easy to bolt together the wrong panels. If you follow the instructions and refer to the numbered panels there will be no major problems.

A concrete footing, or footer, as it is called is required for most multi-plate arches. This footer is usually laid on bedrock, but hard-packed gravel is sometimes allowed when bedrock is not encountered at the designed foundation elevation. A level and transit should be used to install each footer to the same elevation and perfectly parallel to the other. Installing the footers correctly will save many hours of work later on when the plates are bolted together. A footer that is not perfect will cause the plates to bind and the holes will be misaligned.

Concrete footers are constructed like those for a bridge or house. A trench is excavated to the designed elevation, leveled, and then well-greased plywood forms are built to the dimensions shown on the contract plans. All wooden forms should be greased so the concrete will not stick to them. Reinforcement bars are then wired inside the forms as shown on the plans. A water pump may be necessary for each form to keep water seepage out of the foundation trench while the concrete is being poured. The engineer will check for these pumps so it is best to bring the pumps along and expect to use them. A local concrete plant is the best source for obtaining small quantities of concrete required for the footers, but remember to specify the concrete and aggregate mixture that is called for in the contract.

Multi-plate structures should be assembled with as few bolts as possible until all plates are in place. Three or four untightened bolts placed along the horizontal seams are sufficient. When several rings of plates are installed loosely, the rest of the bolts can be installed (also loosely) by working from the center of each panel outward. A bar is often necessary to align these bolts.

When all panels are loosely in place, begin tightening the bolts progressively at one end of the structure. The obvious reason for

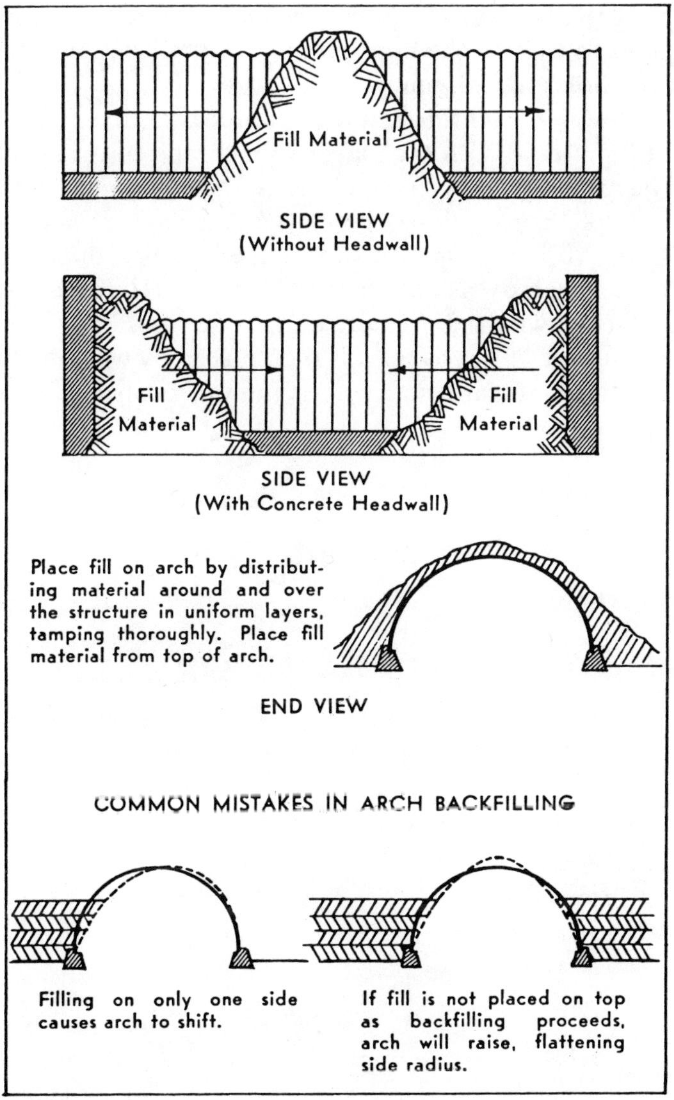

Fig. 6-7. Arch pipe back filling is different than for circular pipes. Care must be taken to avoid shifting and peaking.

this is that the holes in the plate never align exactly. A contractor who begins to tighten the bolts immediately will run into the problem of the holes not aligning.

The best way to backfill a multi-plate arch is to cover it with layers of backfill material with each layer conforming to the shape of the pipe. The arches have a tendency to shift sideways or "peak up" if backfilled like conventional pipes (see Fig. 6-7, lower portion). Start the backfill at the middle of the pipe (Fig. 6-7, upper portion), and keep it as narrow as possible until the top of the arch is reached. This will put downward pressure on the pipe and keep it from distorting. The rest of the backfill is then placed by working from the middle outward toward both ends simultaneously.

7
Finishing

The earthwork operation leaves the subgrade area in rough shape. The expensive excavating machines cannot afford to waste precious time cutting and filling a few tenths of a foot, so when the roadbed is roughly within a half foot of the designed grade, the big machines move on after more paydirt. What is left is a rough, uneven surface that is barely driveable.

The steps that remain in finishing the subgrade consist of grader-blading the rough subgrade to finished design elevations, constructing drainage ditches, finishing culvert catch basins, blading large rocks out of the subgrade, and pulling in overbuilt fill shoulders. When the subgrade is off the design grade by a half foot or more, hauling borrow or waste by truck may be necessary. Despite these many steps don't expect to find finishing as a bid item. It is always a subsidiary item, meaning that it is part of the payment for the earthwork.

Subgrade finishing should be done by experienced personnel, especially grader operators. A grader is easy to run up and down a road, but many years go into mastering the technique of keeping the blade correctly angled while the grader bumps and dips over uneven ground. The tedious work of cutting or filling to an accuracy of a tenth of a foot for days requires patience born from years of experience. The type of grader used for subgrade finishing is shown in Figure 7-1.

Fig. 7-1. A grader is used for subgrade finishing. It takes patience and years of experience to cut and fill a tenth of a foot over rough terrain.

FINISHING THE SUBGRADE

There is no quick, easy way to finish a subgrade. It is impossible to make one grader-pass over rough ground and come away with a smooth, rock-free running surface. Instead, it takes many passes with the grader sifting and pushing aside rocks to create a smooth driving surface.

The standard procedure for working the rough grade is for the grader to make several passes over a mile-long section to obtain a relatively smooth surface. When this is done, the grader goes back over the same area and begins constructing a ditch. This is slow, tedious work. The excavation from these ditches is left in a berm in the middle of the road for future use.

When the ditch is constructed on the slope to the required depth, the grader again blades the road. The object now is to remove all large rocks from the subgrade (usually rocks larger than 3 inches must be removed). To the layman, this appears to be an impossible assignment in rocky soils, yet an experienced grader operator does it all the time.

The grader begins by spreading out the ditch berm across the road and going over it with his blade about 3 inches above the dirt. This technique catches all the larger rock, and the grader periodically sidecasts them as the blade fills up. On the next pass, the blade is lowered to about 2 inches, and on the third pass the blade is skimming the dirt. In this way, the rocks are removed, but the valuable fine dirt remains. In its final pass, the grader "tight blades" the subgrade. The blade is placed right on the ground and scrapes lightly to fill in small depressions. When a section of road is completed, ditches and shoulders should be sharply defined, or the engineer may order the road refinished. With one mile of subgrade smoothed, the grader moves on to the next one mile section.

FINISHING ROCKY ROADBEDS

Any section of roadway constructed with rock presents a problem for the finishing crew because it is impossible to remove all the rock from the subgrade. The standard procedure in such areas is to have a grid roller pulled over the rocky roadway until all of the larger and protruding rocks are crushed or flattened to create a smooth rock floor. To accomplish this sometimes takes several passes with the grid roller. The heavy steel mesh construction of the grid roller is made for compacting and pulverizing rock.

The subgrade is now relatively smooth, but it is still not acceptable, since the spec usually calls for the top 4 inches of subgrade to be free of large rock. Also, since the rocky areas cannot be worked, there will be rolls in the grade to be corrected. The only way to finish the road is to haul in dirt. This is called *cushion material.* This fine dirt must be hauled in from another area and spread to a depth of one-half foot over the subgrade. When the grader is working this material, the operator must be careful to avoid getting the blade down too deep, or it will turn up rocks from the subgrade area.

Cushioning material has been the cause of many court cases because there is no payment for it. The engineer contends that it is covered in the specs, i.e., that the top 4 inches must be free of rocks.

The contractor argues that it is extra work for which he should be paid. A rule of thumb is to look at the extra work that must be done to accomplish this rock-free layer. If the rocky areas are few and small, it is better to haul in a few loads of dirt and get it over with, but if there are miles of rocky subgrade, then some sort of compensation for this extra work should be made.

Often times, the designed quantity of cut does not build the fill. The fill may be built to grade, yet the cut is still a foot above grade, or the cut may be excavated to grade, yet the fill is still a foot below grade. The earthwork crew has no choice but to leave this unfinished road section for the finishing crew to correct.

The engineer can be consulted and occasionally he will allow the grade in an area to be off a little if the road does not suffer as a result. But on high precision projects such as four-lane freeways, the contractor will be required to construct the road to the exact designed grades.

When a road section is below grade, borrow material must be hauled in. The engineer will designate a borrow site, and compensation must be made to the contractor for the clearing, excavation, and haul. When a road section is above grade, the excess material must be removed. Here again, the engineer will designate a waste site. However, no compensation is paid for waste. The foreman of the finishing crew can save many dollars by keeping ahead of the crew for borrow, waste, and rocky subgrade areas. For example, if a road section must have 500 yards of waste removed, it can be hauled to an area which has a rocky subgrade. An efficient finishing crew can do the work twice as fast, for twice the profit.

CULVERT CATCH BASINS

A grader cannot finish culvert catch basins because most are too deep (3 to 4 feet). A small backhoe should be used to shape the basin, and laborers should follow the machine and finish the work with shovels. The backhoe operator should make sure that the narrower lead-in ditch widens gradually into the water catch basin, and that there is a ditch dike on the lower side of the pipe to keep the water from running down the ditch instead of through the pipe. Culvert catch basins are scrutinized by the engineer. He should inspect the first few catch basins while they are being worked in order to avoid doing them over because they do not meet the specification.

OVERBUILT ROADS

The trend today is toward narrower roads, or at least roads which are not wider than specified. A road which is designed to have a 30-foot subgrade may end up 33 feet wide. The engineer may then require the contractor to pull the overbuilt shoulder back to the designed width. A grader with a blade that fully extends and then tips down is the ideal machine for shaving shoulders. A grader with a stationary blade cannot slice off the shoulder because it can not work too close to the soft shoulder. If the engineer requires all overbuilt material to be hauled away, a hydraulic hoe and trucks should be used. Usually this is not done, however, because a 3-foot overbuilt shoulder does not add up to a great amount of material.

8 Bluetops

Bluetops are wooden stakes pounded into the subgrade to give the contractor and engineer a visible reference of the finished grade elevation. Once the stakes are in place, they are painted blue for visibility, hence the name "bluetop." Bluetopping is another form of finishing that comes after the subgrading is completed. This is the final step before the pavement is laid on a road and therefore, elevations must be precise to within a few hundredths of a foot. Figure 8-1 is a close-up view of a bluetop.

Bluetopping is done in conjunction with a base course of gravel. A layer of gravel is first laid on the finished subgrade, and then the bluetops are put in at the shoulders and centerline at 50-foot intervals along the length of the road. Since the gravel is laid on a flat subgrade, the usual pattern is that the bluetops on one side of the road are buried a few tenths of a foot into the gravel, while the other side of the road has bluetops which stick up a few tenths of a foot. The grader must carefully slice off the gravel which covers the bluetops on one side, and deposit the gravel on the other side where the bluetops stick up. This creates a slope for the finished road to drain surface water and to counteract the centrifugal force in automobiles rounding a curve as described below under superelevated curves.

A road is designed from stationary points (large nails embedded into the base of a tree every thousand feet). These are called *terminal bench marks* and serve as references for the designed elevations of

Fig. 8-1. A bluetop is a short, thick stake which is pounded into the subgrade to finished grade elevation and painted blue on top for visibility. A lath is also sometimes used for location purposes.

the road. The bluetopping crew, employing a level and rod, use these bench mark elevations to install the bluetops. After a thousand feet, the crew switches from one bench mark to the next so that the accuracy will not be lost.

SUPERELEVATED CURVES

The centrifugal force acting on a vehicle as it rounds a curve forces it in a tangent line instead of around the curve. As a result curves are built up on the outside to counteract these forces. This build up, called *superelevation*, depends on the steepness of the grade, the degree of curvature, and the speed of the vehicle. On some sharp superelevated curves, the difference in elevation between the shoulders may be as much as 2 feet. When bluetopping superelevated curves, the stakes are pounded into the gravel at the shoulders at 50-foot intervals which take into account the variation in elevation because of the superelevated curve. On sharp curves the distance is reduced to 25 feet.

Someone should be with the grader at all times to inform the operator how far the blade must cut to scrape down to the lower superelevated bluetop. This is not accomplished in one pass. Instead, it takes about three passes to get down to the bluetop level. This gravel is left in a berm in the middle of the road. Care must be taken to avoid knocking out the bluetop stakes, sometimes a penalty is charged when this happens. Once the lower superelevated bluetops are found, the gravel left in the berm can be used to fill in the side of the road where the bluetops stick up. Sometimes there is not enough gravel to build up this side, and gravel trucks should be available to bring in more.

CROWNED ROADS

On a normal straight or tangent section of road, there is no need for superelevation as no centrifugal forces act on a vehicle. When roads are wet, though, vehicles traveling at high speeds actually leave the road and ride on the surface water. This phenomenon, called *waterplaning*, is the cause of many accidents. For this reason, a paved road is never designed flat, but with a slight outslope to the paved surface. This sloping is called a *crown*. The crown is highest at the centerline of the road and slopes out toward both shoulders so surface water can run off. The usual slope rate of the crown is 3%, which means three-tenths of a foot drop for every 10 feet. The slope of the road in crown sections is not nearly as severe as with superelevations, and the bluetops are simpler to find with the grader blades. Usually, the shoulder bluetops at each side of the road are buried a few tenths of a foot, and the centerline bluetop is raised a tenth or two. The grader works both shoulder bluetops first until they are scraped down, leaving any excess material in a slight berm in the center of the road. The grader then completes the job by spreading this gravel over the centerline area until a slight apex, or crown effect appears.

Here again, an operator can't be in a rush when finishing with bluetops. It is a slow and exacting process. It is best to work a half mile of bluetops at a time, or the operator will tend to forget which sections are too high and which are too low. Bluetopping sounds like a difficult job, but it is not when an experienced grader operator is at the controls. Remember, the uneven, rolling roadbed has already been smoothed by the finishing operation, and the grader need only find the bluetops to do an acceptable job.

After the bluetops are found and the road is shaped, a steel drum roller is used to pack down the loose gravel. Watering is also necessary in order for the roller to obtain a hard running surface. Once this is done, the gravel will keep the crowned and superelevated slopes until paving time.

9
Rock Crushing

The production of aggregate (crushed rock) is one of the most expensive items of a contract. The rock-crushing operation is a separate project in itself, and is so big and complicated that most road contractors subcontract their crushing needs to another contractor who does nothing else but crush rock.

HOW A CRUSHER WORKS

Rock is crushed in a machine called a *crusher*, though this single term belies its actual appearance. A crusher is actually several interconnected machines which do different crushing, sorting, and sieving jobs simultaneously. The product is a specified gradation of gravel that moves off the output feeder belt to the waiting trucks.

Electrical power is used to operate the many electrical motors that run the belts, jaws, rolls, and cone of a crusher. A main control switchboard in the generator trailer can quickly stop a conveyor belt or cone that suddenly malfunctions. The generator is located in a large trailer for easy mobility.

All crushers have a feed-in funnel into which the loose rock is deposited. In some larger crushers this feed-in funnel has a conveyor belt which can be controlled by the operator to feed just the right amount of rock into the crusher. The advantage of the feed-in belt is that large amounts of rock can be piled at the feed-in funnel, without the operator worrying about the crusher running out of rock. On smaller crushers, the feed-in funnel leads directly to the primary rock-crushing jaws, and it must be fed constantly by another machine. In a pit where a dozer is pushing rock directly into the crusher, the direct feed-in funnel is used.

After the rock is fed into the crusher, it enters the jaws of the crusher, a large, oblong mass of steel that rocks back and forth rapidly, crushing the large rocks into smaller ones approximately 6 inches in diameter. The other smaller rocks and dirt fall right through. Large crushers can run 30-inch rocks through their jaws, but such large rocks should be kept to a minimum because of the time required to reduce their size. Maintenance is also a problem because of the power necessary to break up large rocks. Crushers that handle jobs of 60,000 to 100,000 tons try to keep the rock under 24 inches, and the smaller, 30,000 to 60,000 ton crushers try to keep the rock under 18 inches.

Once the rock passes the primary jaws, it is 6 inches or less. This material is then screened to sieve out all the small rock and dirt. The remaining oversize rock runs through circular, solid steel rolls which turn rapidly and crush the 6-inch rocks to half size. This material is then run across another screen and the small material drops through. The remaining oversize rock is run through the cone, a cone-shaped piece of steel that fits inside an outer collar and vibrates rapidly to crush the remaining rock against the collar. The cone is set to produce gravel at the maximum allowable size called for in the specification. For instance, ¾ inch gravel has the cone set at ¾ inch to begin with. Figure 9-1 shows a crusher in operation. The pile of rocks on the right are fed into the crusher jaws. The crushed rock is transported across a vibrating screen (center) to a cone and steel rolls that reduce the rock size to the finished size shown on the left.

When a crusher starts up, it is not uncommon to produce gravel that does not meet the specs. This happens because the dimension at which the machine was set changed because of vibrations during transport. The jaw, drum, and cone can be adjusted to change the size of rock, and sometimes it is a simple matter of changing the size of a screen to get the rock to specified size.

The size and complexity of a crusher depend on the design requirements of the aggregate. The base course of gravel used for bluetopping, or for county roads, is well graded with about 10 to

Fig. 9-1. A crusher in operation: The works at the right feeds the piled-up rocks into the crusher jaws which crush the rock down to 6-in. size. Then it vibrates across a screen which takes out all the small rock and fines (center). The remaining rock then runs through steel rolls and a cone which further break the rock down. The well-graded finished product is shown at the left being fed into the waiting bin for the trucks to haul it away.

15% fines (dirt). This type of material is easy to produce, and often all that is needed is a jaw and rolls.

Aggregate for pavement is a different matter. Fine dirt and rock fines must be removed from the gravel by the crusher. This necessitates the use of a reject belt. The only thing that comes off this belt is dirt and rock dust, and it is hauled off to a waste site designated by the engineer. Handling reject material is an extra expense that should be taken into account when bidding a job. When the specifications call for fines to be in the 0 to 4% range, expect to use a reject belt.

The setup for the reject belt comprises another series of screens which sift out the fines from the material and onto the reject belt. The belt carries the rejected material from the crusher and dumps it into a truck. In the average rock source, expect one truck of reject for every four trucks of aggregate.

SETTING UP A CRUSHER

Proper setting up of a crusher is critical to its efficient operation. The contractor must look at the rock source and determine the best place to locate the feed-in funnel of the crusher. Ideally, this funnel should be butted up against the rock source so that a dozer can push the loose rock from the quarry or pit right into the jaws of the crusher. This is not always possible, however, and the rock must be trucked to the crusher site. Trucking incurs an added expense which should be included in the bid.

The separation of a crusher from its rock source is most often the result of not enough room at the rock source site. Rock sources are often outcrops along a narrow stream channel where there is no room to set up a large crusher, so the crusher site must be located at the nearest suitable area for setup. A naturally flat area is best for a crusher setup, but this condition rarely occurs. Instead, the contractor must create a flat spot with his dozer, an operation that is called *crusher development.*

An acre of land is the minimum area necessary to set up a crusher. Sometimes two acres are needed for larger crushers. One of the first things that a contractor should look for when bidding a crushing job is whether the designed crusher site is large enough to accommodate an efficient rock crushing operation. Too small a site has resulted in trouble for many contractors, a mistake easily made because a crushing operation needs much more room than the space occupied by the crushing machine. Space is needed for the large generator trailers that power the crusher, for loading the gravel trucks, whether leading straight from the belt or from an endloader, and space is needed for a good sized stockpile site so that if the trucks fall behind the crushing operation, the crusher will not have to shut down until the trucks catch up. Figure 9-2 shows the pile up of gravel from an output feeder as the result of a truck holdup. Figure 9-3 is a photograph of a stockpile at a crushing site.

The development of the crusher site and the pit site should not be

Fig. 9-2. The result of a truck hold-up. Gravel from the output feeder belt pours onto the ground and a loader must remove it. This is a common occurrence.

Fig. 9-3. Sometimes rock must be crushed and stockpiled because no trucks are available, or the road is not ready for it. A stockpile site should be considered when bidding a crushing project.

at the contractor's expense. Compensation should be paid for the pit or quarry site for clearing, pit development, drilling, and blasting. The contract should compensate for such crusher site expenses as clearing and all excavation required to obtain the designed flat area. This item is often overlooked in a contract, even though crusher site excavation may run as much as several thousand cubic yards.

The development of the pit and the crusher sites is sometimes a subsidiary item, and the contractor must add the cost of the development to the cost of the gravel. A contractor can easily estimate most development costs by estimating time and equipment to do the job. However, drilling and blasting costs are more difficult to determine. A good rule of thumb is to charge $1.50 for drilling and blasting each yard of hard rock up to 30,000 cubic yards and $1.00 per yard thereafter.

THE ROCK SOURCE

There is a big difference to a crushing contractor whether he is to crush rock that comes from a pit or from a quarry. A pit contains rock in a loose or fractured state and can be excavated by a dozer with little or no drilling and blasting. A quarry, on the other hand, is a solid rock cliff or knob that must be drilled and blasted.

A pit can be developed and prepared even as the crusher is being set up whereas a quarry often takes weeks of pioneering, drilling, and blasting before the crusher can be moved in and set up. Another big problem with a quarry is that it is very difficult to bring equipment to the quarry top so that vertical drilling can be done. Instead, the dozer pioneers until it hits immovable rock. Then a drill must put in 10 feet deep side holes and blast the rock. This operation is repeated over and over again until the top is reached. Much of the equipment is standing around during this tedious operation. A contractor should take a hard look at any proposed quarry and decide how difficult it will be for his equipment to gain access to the top. If special measures must be taken, compensation should be made in the cost of the rock or quarry development to warrant this extra effort.

The hardness of the rock in a quarry should be analyzed because hard rock requires closer hole spacing and the use of more explosives. The crusher also suffers from hard rock, and maintenance problems are sure to result. To check on the properties of the rock in a quarry, consult the contracting office for geological data that gives core drill results, weight of rock per ton, hardness, breakability, etc. The geological data will also show any dirt seams that run through the rock. These dirt seams can cause havoc with drill holes because the dirt can fall in and block them.

A pit contains a large amount of fines, and is not the best rock source for asphalt pavement since it requires a very low amount of fines. On the other hand, a pit is excellent for gravel for use as a bluetopping base course, or for a graveled road. Quarries, too, can cause problems. While they are a good source for crushing when fines are to be kept at a minimum, they cause headaches when gravel must be produced with a large amount of fines. The only way to get fines in a hard-rock quarry is to pulverize some of the rock. This requires additional work and wear and tear on the machine.

A contractor should weigh several items when looking at a rock source—the rock specification should be checked for the degree of difficulty to produce the required products, the crusher site should be analyzed for the ease of establishing the crushing setup, and the pit or quarry should be checked for the particular project.

10 Gravel Surfacing

Hauling gravel from the crusher to the road and distributing it in an even layer would appear to be a simple operation, yet somewhere along the line the process becomes very complicated. It is not merely a matter of dumping crushed rock on a road and then spreading it out. Gravel must be laid on a road to the specified depth, a chore not as simple as it sounds. The rock must then be windrowed, watered, mixed, relaid, and then compacted. A windrow is material that is dumped and then bladed into a long row as shown in Figure 11-2.

VEHICLES FOR HAULING GRAVEL

Gravel is hauled in trucks or belly-dumps. The belly-dump shown in Figure 10-1 is the best machine. It has a low center of gravity which makes it much safer than a truck, and the gravel is disposed of out of the bottom, much like a scraper. The belly-dump can lay down an even thickness of gravel without the gross irregularities of dumping by truck. The problem with belly-dumps is that their design is limited to only gravel hauling. A truck, on the other hand, can be used for any type of hauling.

Spreading gravel by truck is not smooth, and usually, when the bed is first raised a huge pile of gravel is dropped that is hard to drive over. Trucks also are placed in an extremely precarious position when they raise their 30,000 pound load in the bed. A truck in this position should never be driven on an irregular surface because the

Fig. 10-1. The bellydump truck is ideal for gravel hauling because of its low center of gravity, and because the gravel drops out its bottom in a smooth lift.

center of the load could shift and cause it to tip. While trucks are less efficient or safe than a belly-dump for hauling gravel, they are more practical overall, and are therefore, used more frequently. Figure 10-2 shows a grader smoothing out truck piles to reduce the danger of tipping the truck.

Fig. 10-2. A grader should be nearby to periodically smooth out truck piles to avoid the danger of tipping over a truck.

DETERMINING THE PULL OF A HAULING MACHINE

Gravel hauling vehicles must dump their loads in an area of the road that will give the exact rock depth when compacted. If the gravel is spread over too long a distance, the rock depth will be too thin, and if the gravel spread is too short the rock depth will be too thick. Both of these situations will cause the contractor extra work to correct. This is easily avoided by determining the "pull" of the hauling vehicles.

The pull of a haul vehicle is that distance along a road in which the gravel must be dumped in order to get the specified gravel thickness

Road Width in Feet	Truck Capacity in Cu Yds							
	10	12	14	16	18	20	22	24
16	28*	34	40	45	51	57	63	68
18	26	31	36	41	46	51	57	62
20	23	28	33	37	42	47	52	56
22	22	26	30	34	39	43	48	52
24	20	24	28	32	36	40	44	48
26	19	22	26	30	33	37	41	45
28	17	21	24	28	31	35	38	42
30	16	20	23	26	29	33	36	39
32	15	19	22	25	27	31	34	37
34	14	18	21	23	26	29	32	35
36	13	17	19	22	25	28	30	33
38	12.5	16	18	21	24	26	29	31
40	12	15	17	20	23	25	28	30

*Distance in feet.

Note: Add 3 ft to road widths for shoulder tapers. Above pull data is based on a rock depth of 6 in.

$$\text{Pull Distance} = \frac{\text{Truck capacity in cu yds}}{\text{Road width in feet} \times \text{gravel depth in feet}} \times 27$$

Fig. 10-3. Pull chart for gravel surfacing. Always add the gravel taper distance on shoulders to total road width.

after compaction. For example, a 30-foot wide road which is to have a 6-inch layer of rock requires about 1 cubic yard of gravel for every two feet of road length. Therefore, a 10-yard gravel truck must spread its rock along 20 feet of the road in order to get a 6 inch layer of rock. Figure 10-3 shows the pull distances for a wide range of truck capacities and road widths when the depth is 6 inches. For other gravel depths use the formula shown.

If all the vehicles have the same capacity, it is a simple matter to calculate the pull distance and then stake out the road. Each vehicle then drops its load between a set of stakes as the graveling progresses along the road. However, it is not always possible to have vehicles of the same capacity, and a dump man must be on hand to measure the spread of each vehicle.

LAYING DOWN GRAVEL

It is best to begin laying down gravel at the end of the road and work back toward the crusher. The grader can then begin to process the rock without interference from the vehicles hauling back and forth. The mistake is often made in thinking that gravel can be laid and the loaded gravel rigs will compact it as they travel over it. This is not true. Gravel rigs do not create the smooth, compacted surface that a steel drum roller does, and haul rigs cannot drive close to either shoulder for fear of tipping over.

When gravel rigs are spreading, a grader should be on hand to smooth out the humps and bumps caused from irregular pulls. The grader at this point does not attempt to process the gravel. It simply smooths out the rock so that the rigs can drive safely. The usual procedure is for the grader to process a few thousand feet of gravel, go down to where the rigs are spreading, knock down the bumps, and then go back to processing again. This not only saves the use of another grader, but keeps the roller from falling too far behind the processing.

PROCESSING GRAVEL

The first step in processing gravel is to check it for moisture. If it has been raining, the chances are that the gravel is wet enough.

Otherwise, a water truck must wet down the gravel. This is necessary, especially in hot, dry weather, because the gravel quickly dries out and segregates, with the fines sifting to the bottom and the larger rock raveling on top.

The rock is watered and the grader then scrapes it up into a windrow in the center of the road and then pushes it back and forth twice to thoroughly mix the gravel. When the gravel is moist and well mixed, the grader spreads it out to the required width. Thick spots in the gravel depth are usually ignored, but thin spots should have extra gravel dumped in them immediately.

The gravel should be compacted immediately after it has been processed. A self-propelled steel drum roller is most often used. Sometimes the roller falls behind the processing grader, and the gravel begins to dry out before it can be compacted. When this happens, water must again be applied lightly to insure compaction. On days when heavy rains occur, the gravel will become saturated, and it is advisable to cancel processing operations. Figure 10-4 shows the grader making a final pass on the gravel, followed closely by a tractor with a roller.

Fig. 10-4. A grader makes the final pass on the gravel and a tractor with roller works close behind. The rock must be damp to compact the gravel.

11
Asphalt Pavement

Asphalt is a natural ingredient of most petroleums in which it exists. Crude petroleum is refined to separate a wide range of useful products and to recover the asphalt. Asphalt is of particular value in road construction because it is a strong cement, readily adhesive, highly waterproof, and durable. It is a plastic substance with controllable flexibility, and when mixed with aggregate, it forms an asphaltic concrete pavement. Asphalt is also highly resistant to most acids, alkalies, and salts. Although a solid at ordinary temperatures, asphalt may be readily liquefied by heating, or by emulsion in petroleum solvents and water.

There are countless types and mixtures of road paving asphalts, the names of which tend to confuse a contractor. Even experienced engineers sometimes scratch their heads in confusion. Asphalt distributing firms are the best source of information when trying to decipher an asphalt that goes by the name, Cutback RC-MC 30. These firms can simplify the complicated nomenclature, and explain what it is and how it should be used for asphalt paving.

The key to good asphalt pavement construction is the intelligent selection of not only the aggregate, but also the method of paving that will economically and adequately meet the design requirements and weather conditions. There are usually several different methods from which to choose. The control required on the finished pavement should be considered when selecting the process. For example, using a motor grader for spreading when the tolerances must be within a few thousandths of a foot is poor, if not disastrous, judgment. In this instance, a pavement machine, with its many controls, should be used. Grader mixing is alright for paving low standard county or forest roads.

Poor workmanship is responsible for most substandard paving jobs. Attention to small details as well as to proper procedures cannot be overemphasized.

Weather plays a critical part in pavement operations. The base course and seal coat are both highly susceptible to low temperatures when being applied. Paving specifications always call for a minimum temperature to be reached before paving can begin. Rarely is paving allowed when the temperature drops to the low 60's.

A contractor who bids a forest paving job high in the mountains may find August too late in the year to economically pave because the minimum temperature restriction allows paving for only a few hours each day. In some mountainous areas the weather may turn bad in late August and it will rain frequently until September, when it begins to snow. If the job site looks like an area that tends to be cool and wet in the late summer, consult the local weather station. Mid-July is the best time to pave in high-country areas.

PRIME COAT

The subgrade of a road must be prepared so that excessive quantities of moisture will not weaken it and cause the pavement to fail. Moisture—frost—is the worst enemy of an asphalt pavement. By keeping moisture out of the subgrade, the danger of frost heave is at a minimum.

The process by which a subgrade is protected is called *subgrade priming*, or *prime coat*. A prime coat is a low-grade asphalt which is applied to the subgrade surface and seals it so that atmospheric moisture cannot penetrate to the ground below. The prime coat may be applied to a subbase of gravel or directly to the ground.

The prime coat of asphalt is applied with a distributor truck, which is a tanker having both a heater to keep the oil hot and adjustable spray nozzles spaced on bars at the rear. The tanker begins a trial run by spraying a 100-foot section of subgrade. The meter on the tank is then read to see how many gallons were applied to the section. If the application is too heavy, the asphalt will run, and if it is too light the subgrade will still show. The engineer will then have the driver adjust his speed and another test section will be run. When the correct speed and application rate combine to create a satisfactory coating, the rest of the priming job will be easily completed using this standard. Weather and subgrade conditions may vary and another speed may have to be determined as the job goes on, but this does not hold up overall production. Both the contractor and engineer know how much asphalt is being applied for every square yard of road subgrade. This will vary from 0.25 to 0.45 gallons per square yard with the usual rate about 0.33 gallons per square yard. A distributor truck is shown in Figure 11-1.

The subgrade must be moistened before the prime coat is applied. Be careful, for too little water makes it impossible for the asphalt to penetrate and seal the subgrade while too much water will halt penetration of the asphalt by displacing it in the subgrade. Both of these conditions leave the asphalt on the subgrade surface in

Fig. 11-1. Asphalt prime coat is applied with a distributor truck which sprays a constant thickness of hot asphalt through a series of nozzles.

globules, and the engineer will order it scraped off to the side and another application begun.

If traffic is permitted to drive over this prime coat, even after it has penetrated and dried, its seal will soon be broken, and the entire priming process wasted. The gripping action from the traction of vehicle tires simply tears apart this light asphalt layer. Since it is usually impossible to keep traffic off a road for weeks at a time, blotter material is placed over the prime coat.

A blotter is fine sand or gravel which blots the excess asphalt and protects it from vehicular traffic. The most widely used blotter material is the reject fines from the crusher. The priming asphalt is usually given about 5 minutes to soak into the subgrade before the blotting begins, otherwise the blotter will soak up the asphalt before it penetrates the subgrade.

A spreader box should always be used for spreading blotter material. All that is needed is about a 1-inch layer of blotter material to protect the prime. Having a truck spread it would be disastrous. A spreader box can be connected to the rear of a truck, or it can be a self-propelled machine which has its hopper fed from the rear by a truck while the blotter is spread from the front so that its tires will not break the prime seal. The spreader box also has a screen to catch oversize rocks that could punch holes in the prime seal. Spots on the blotter which show bleeding should have more blotter added. This is usually done by a laborer who drives along behind the spreader box with a half yard of blotter in the bed of his pickup.

It is possible to drive on the blotter half an hour after it has been applied without compacting it, but there is less chance of an accident if a roller first makes a pass over it. Blotted prime coats hold up for weeks to normal traffic prior to laying of the base course of asphalt.

BASE COURSE

The asphalt base course can be manufactured in a mixing plant set up at a central location on the project, or it can be mixed on the road. Mix plants are used when large quantities are required and a paving machine must be used. Road-mix pavement is used for low volume, low standard roads.

Road-Mix Pavement

Road-mix pavement (also called mixed-in-place) is suitable for all types and loads of traffic when the road is not as critical as a freeway. Alternate truck routes and secondary state highways are two types of roads which utilize road mix. The principal advantage of road mix is that very little equipment is needed. Dump trucks deposit the aggregate on the road quickly since there is no mixing at a plant to slow them down. Therefore, the paving process goes quickly.

Traveling Mix Plant

Road mix may be applied using a traveling mix plant which moves along the road, picks up the windrowed aggregate, adds asphalt to it, and then leaves the pavement in a windrow behind it. However, further mixing of the pavement is often necessary. One or more motor graders continue to mix the windrowed pavement until all the aggregate is coated with asphalt. This coating process normally requires three quick grader passes. When a long enough section of windrow is ready to be applied to the road, the graders simply spread it out to the depth and width specified, but not in lifts that exceed 4 inches. Therefore, if a pavement depth of 6 inches is required, it should be applied in two, 3-inch lifts, and each lift should be rolled separately with a rubber-tired roller. A maximum lift of 3 inches is necessary since the type of asphalt used for road mix pavement will not bind the aggregate adequately in thicker lifts.

Another type of traveling mix plant adds the asphalt after the aggregate is deposited from trucks into its feeder bin. The plant mixes and then lays down the pavement to the width and depth required. Care should be taken to make sure that enough of the asphalt solvent has evaporated to make the mix set up when the roller passes over it. Unfortunately, too much solvent may evaporate from the top surface to the extent that it will not bind, while solvent remaining in the inner mix cannot escape thus preventing the inner mix from setting up. The end result will be a parched and rubbery pavement which will soon be torn up by traffic. To alleviate this, graders scrape the mix into windrows and continue to mix it until it is aerated enough to lose most of its solvent.

Road Mix With Grader

Grader mixing is the fastest and most economical method of paving. Aggregate is laid in a flattened windrow in the middle of the road and an asphalt distributor truck applies the asphalt in successive passes. Between passes the grader turns over the windrow to expose more uncoated asphalt, and continues to mix it until all the aggregate is coated with asphalt. The grader then spreads the mix on the road in 3-inch lifts, and a rubber-tired roller completes the pavement operation. Figure 11-2 shows a road-mixed windrow of asphalt aggregate.

Fig. 11-2. Road mixed asphalt-aggregate is windrowed on the subgrade. A grader will now spread it out and a roller will compact it.

Mixing Plant and Paving Machine Method

The plant mix and paving machine method produces the highest quality paving job. Both paver and mix plant are calibrated to produce and apply pavement to the most rigid standards. Though this method is slower and more expensive, the resulting pavement will be of freeway job quality.

Because of the superior quality of asphalt produced in a hot mix plant, this is the most widely used method when paving to rigid specifications and road standards. There are several advantages to a hot mix plant:

1. Aggregate can be accurately proportioned, dried, heated, and mixed so that all particles are completely coated with a film of asphalt.
2. Mix plants are portable and can be transported to a central location on the project.
3. Mixing is independent of weather because the aggregate can be dried immediately after the rainfall ends and paving begun soon thereafter.
4. The amount of asphalt can be accurately controlled and heated to provide the proper viscosity for thorough mixing.

In order to produce the highest quality asphalt plant mix, uniformity and balance are essential. Of great importance are the uniform quality and quantity of the material being fed into the plant. Before beginning production, the asphalt storage tank should be full, and maintained in as near a full condition as required to keep the plant in continuous operation. Due allowance should be made for delays in asphalt delivery to the storage tank. Also, there should be an adequate stockpile of aggregate to keep the plant running full at all times. The mix plant is a delicate machine and inadequate quantities of either asphalt or aggregate will produce a pavement that is inferior.

The mixing process begins when the trucks dump their loads into feeder bins. The aggregate then enters the dryer and a controlled flow of air and hot gas heats the tumbling aggregate to the correct temperature for asphalt binding. The aggregate is then fed into the waiting bin. As it is needed, the aggregate is fed into a power mixer called the *pugmill* where it is sprayed with asphalt as it is tumbled. When the asphalt is completely mixed, it moves to another bin where it can be dropped into waiting trucks to be hauled to the paving machine. When the mix plant is in full operation, the aggregate should be moving continuously through the plant, assuring a high output of asphalt mix.

Cold mix plants are simpler than hot mix plants because the asphalt and aggregate are not heated. However, the asphalt does require the addition of a solvent to make it soluble. This is the most critical phase of the cold-mix production process. Aggregate is dumped into a hopper and is fed directly to the pugmill section where the tumbling aggregate is mixed with rapid-curing asphalt and then mixed with an emulsion of water and solvent. This mixing process is much more delicate than that of the hot mix because the correct emulsion is essential to the quality of the pavement. The asphalt supplier will furnish the correct data for proportioning the asphalt and solvent.

After the mix is completed it is loaded into trucks and hauled to the paving machine. The cold mix sets up after the water has evaporated, taking with it the solvent and leaving the asphalt to return to its original semisolid state. Cold mix pavement sets up slowly compared to hot mix, and it should not be driven on for at least 24 hours after application. Usually one lane is paved at a time and the opposite lane is used for traffic directed through by a slow moving pilot car. Although cold mix pavements are not used as much as hot mixed, their use has increased substantially in recent years.

A paving machine can be used to apply either hot or cold mixed asphalt pavement. The paver is hooked to a truck loaded with mix and is fed by the truck as it moves continuously along the road. The paver consists of a feed-in hopper, conveyor, distributing screws, and screed. The pavement can be laid accurately to tolerances of thousandths of a foot with a paver. The paving machine should move along the road at a constant, determined speed. If a paver has to stop for any length of time, the mix cools and loses its optimum viscosity, resulting in roughness and raveling of the finished surface and poor compaction. Therefore, the speed of the paver should be adjusted to the output of the plant and capacities of the trucks to insure continous operation.

Pavement joints are a continous problem for the paving machine. When a road is paved one lane at a time, which most are, there will be a joint between lanes with which to contend. There are several rules to minimize the difficulty of joint binding:

1. Joints on succeeding layers of pavement should never be directly above the other. The overlayer joint should be offset about 6 inches from the underlayer joint.
2. Keep joints straight. It is impossible to get a smooth joint if the paving machine is wandering back and forth.
3. Maintain as nearly vertical a joint on the first lane as possible.
4. Keep joints uniformly overlapped by about 6 inches.
5. The joint should be rolled as soon as possible after it is laid, and the roller should run on the previously paved lane with only the outer edges of its wheels on the joint. This places maximum pressure on the humped-up joint and squashes it down to road level.

Rolling can be done with either a pneumatic rubber-tired roller or a steel-wheeled roller. A steel-wheeled, self-propelled roller is shown in Figure 11-3. Some Engineers prefer the rubber-tired roller because the kneading action of the tires tends to lock the surface particles together and create a traffic seal. Two rollers should be used, one for rolling the joints and the other for rolling the rest of the newly applied pavement. Rollers should move slowly but uniformly. The speed of a steel-wheeled roller should not exceed 3 miles per hour, while rubber-tired rollers should not operate at more than 5 miles per hour. The line of rolling should not be suddenly changed, nor should the direction be suddenly reversed. This will cause displacement of the mix. Any pronounced change in direction of a roller should be made on stable material. If rolling causes any displacement of pavement, the affected area should immediately be loosened up with stiff rakes and restored to the original loose grade before being rolled again. Rollers and other heavy equipment should not be allowed to stand on a finished paved

Fig. 11-3. A steel-wheeled self-propelled roller produces a smooth, flat, compacted surface. Note the pilot car which regulates traffic flow and speed.

surface before it has thoroughly cooled or a permanent depression will result. After the first rolling of the hot mix, another rolling should be made while it is still warm because the pavement tends to spring up a little after the first rolling.

SURFACE COURSE

Asphalt surface treatment is a broad term covering several types of asphalt and asphalt-aggregate applications, usually less than 1 inch thick. The types of surface treatment range from single, light applications of liquid asphalt to multiple surface courses made up of alternate applications of asphalt and aggregate. The purpose of these treatments is to seal and add life to the road surface. A surface treatment is not a pavement in itself, even though it may have aggregate in it. Instead, it provides waterproofing and resistance to abrasion from traffic. Since surface courses are thin, as noted above, they add very little to road strength.

The most commonly used surface treatment, or seal coat, is the asphalt-aggregate seal coat. This greatly improves the texture of the base course. The aggregate is referred to as CHIPS and is usually ⅜ inch or smaller in size with very little fines. These small chips knit in between the larger aggregate of the base course to make a smooth riding surface.

The asphalt distributor plays a critical role in the surface treatment of a road. It is essential that the distributor be able to spray the asphalt uniformly over the base course. To do this, all spray nozzles must be maintained at identical pressure and temperature. The spray bar must also be set at the height suggested by the manufacturer to provide complete and uniform overlap of adjacent sprays. The speed of the distributor is also critical. Once the correct speed has been determined, it should be maintained at all times while the distributor is spraying. Daily inspection of the distributor is essential to insure a good job. Many seal coats have been ruined because one or more spray nozzles came out of adjustment halfway through the job.

Spreading Aggregate

Aggregate spreaders vary from a simple gate box attached to the back of a dump truck, to a very efficient self-propelled unit which applies larger size chips on the bottom and smaller size on top. A self-propelled unit appears in Figure 11-4. The self-propelled spreaders are by far the most widely used for seal coating because of the accuracy with which they control the thickness of the chips and the width of their spread. Before the distributor begins spraying, the aggregate spreader should be filled and ready to begin applying aggregate right behind the distributor. A common fault is to operate the distributor too far ahead of the spreader.

Both pneumatic rubber-tired and steel-wheeled rollers can be used to compact surface treatments. Rubber-tired rollers, however, are preferred because steel wheels tend to fracture the chips.

Traffic should be kept to a minimum during the seal coating to prevent loss of the chips before they are cemented to the asphalt. A pilot car should always lead the traffic through at a very low rate of speed.

Fig. 11-4. A self-propelled aggregate spreader spreading chips for a seal coat. Note the trucks in the background waiting to dump aggregate into the spreader's hopper.

Fog Seal

The seal coat chips become embedded into the asphalt until all the asphalt surface has chips cemented to it. There are, however, loose chips remaining on the top that cannot be cemented in because the asphalt is unable to accept more rock. These loose chips pose a danger to vehicle traffic because of skidding. A self-propelled broom is used to whisk any remaining loose chips off to the side of the road, leaving behind only the chips that are cemented into the asphalt coating. Then a fog seal is applied to the surface at a rate of 0.1 or 0.2 gallons per square yard to further embed the chips. This sprayed-on solvent darkens the road surface to make it more visable in fog, thus the name *fog seal.*

12 Riprap

Riprap is coarse rock fill used for erosion prevention. It varies in size from 6 inch size to 5-ton boulders. Riprap in the 1 ton class does not consist of only 1-ton rocks. Instead, there is a gradation starting at about the foot size and ranging up to 1 ton. These smaller rocks are necessary to create a binding between the large rocks and to fill in the voids.

Riprap is often a work item on road and bridge contracts. It is used on roads to protect fill slopes from nearby streams, and on bridges to protect the abutments from flooding and ice flows. There are also contracts concerned only with the production and placement of riprap. Where flood-swollen drainages have eroded roads, riprap projects of between $100,000 and $500,000 are not uncommon. Some contractors work almost exclusively on riprap contracts.

Bidding a riprap job is quite different from a road job. A road job has many different pay items, and the cost of each item is fairly well standardized. Riprap projects have only a few pay items and their cost varies greatly because of differences in the rock source, length of haul, and difficulty of placement. Only a man who is experienced or who has been carefully tutored in riprap contracts should bid a riprap job. A person without the critical eye for this type of work may underbid the actual cost of the work.

THE RIPRAP SOURCE

Unlike the contractor who when blasting for road construction cares little about the size of the shot rock, the contractor shooting a quarry for riprap is concerned about how the rock is going to break. Rocks that break too large for the specified riprap must be drilled and shot again. Reblasting the boulders usually disintegrates them into a pile of gravel. Rock that breaks too small is also a waste.

The hardness or softness of the rock at the quarry and its fractured seams are the principal factors in successful riprap production. The perfect drill pattern and explosive load are wasted if they do not correspond to the hardness of the rock, and the rock will end up too large or too small. On riprap projects that require gradations ranging from 12 to 48 inches, errors in riprap size are less critical because of the great size range. On projects where the riprap is all one size, expect some waste of the larger and smaller rock, and make sure the quarry site is large enough for this material to be stockpiled.

When riprap breaks too large discuss the matter with the engineer and volunteer the larger rock at no extra cost. Riprap is priced according to size, the larger riprap being more expensive.

There is nothing more disturbing to a superintendent than to go to the care of producing perfect sized riprap, and then have it rejected because of too many fines (dirt). Most contracts call for 0 to 5% fines in the riprap. Now that's clean rock! Even in solid rock quarries it is difficult to produce riprap that clean. The end loader can pick up that much dirt from the pit floor during the loading operation.

A new quarry with heavy overburden (a dirt layer over the rock) or with dirt seams will certainly have many fines to contend with. When this occurs the contractor should use a rock rake on the quarry dozer. A rock rake is a dozer blade with 1-foot openings spaced across the blade, and teeth that extend downward. The rock rake pushes the mixed rock and dirt a few feet at a time and keeps raising the blade to turn over the rock. This "combing" process agitates the rock so that the fines sift through the holes in the blade and the voids in the larger rock. After combing dirty rock for about

30 feet, the product will be clean rock with only a few percent fines.

The front end loader in the pit should be a heavy-duty machine. If riprap in the 18-24 inch range is to be loaded, a loader in the 170 horsepower range is desirable. Larger rock, up to 2 tons, requires a loader of 325 horsepower. Loader tires are extremely susceptible to sharp rocks, and at a price of several thousand dollars per tire, it doesn't take many ruined tires to cut deeply into a contractor's profits. There are steel mesh nets that fit over the tires on a loader (like tire chains) to protect them from rocks. On a large riprap project nets pay for themselves many times over. A steel mesh net over a loader's tire is shown in Figure 12-1.

Fig. 12-1. Steel mesh nets which fit on a loader's tires like chains will save thousands of dollars when working in a rocky area.

Another handy item for removing dirt from riprap is a "grizzly," or "slat bucket" for the end loader. This is a bucket with teeth and vertical steel slats, separated so that fines will sift through the openings. Remember though, the tires on the loader will be more vulnerable to the smaller, sharp rocks which fall through the openings.

RIPRAP HAUL TRUCKS

The ideal truck for hauling riprap is a heavy-duty, 18-cubic yard dump truck with a solid steel dump bed. This bed will take the extreme punishment of thousands of tons of large riprap. Contractors who do riprap work exclusively use this large sized truck. Dump trucks in the 12 to 16 yard range without heavy-duty construction are slowly worn down by the steady pounding of rock. These smaller trucks are made for much lighter duty service such as dirt or gravel hauling. Economics, however, often dictate the use of equipment, and a contractor is forced to use light-duty trucks. In order to protect the trucks as much as possible, the operator must be extremely careful when loading. A difference of 1 foot in the drop of the riprap can add many extra hours to the life of a truck. As mentioned before, if a little dirt is dropped in the bed of the truck, it will act as a cushion when the heavy rock pounds into it.

Wood planking bolted to the bed will also improve the situation, but the wood tends to gather the wet fines until much of the bed is a mass of compacted mud stuck to the planks. Another good idea is to bolt or weld small gauge railroad rails running parallel to the length of the bed and spaced about 1 foot apart. These not only withstand punishment, but also cut down on the hold-back friction, a major cause of trucks tipping over.

DUMP MAN

A dump man is imperative when trucks are end-dumping over a road shoulder into an eroded area. His tasks include keeping track of the quantity dumped into each work area and informing the truck drivers where to dump in each area. He also backs them up to the edge of the road and throws a block under the rear tires, a practice that has saved many trucks and many lives.

A truck must sometimes raise its dump bed loaded with riprap even though it is not on a flat surface. The dump man stands back and watches the bed and warns the driver if the load shifts. Such a shift can cause tipping. The problem is created by the wedging of riprap rock against the sides of the bed. When a bed is half-raised, rock should begin sliding out of the truck bed, thereby relieving much of the weight problem. However, when the rock gets stuck, the bed raises up and sways precariously. If the truck is on a slope, it is easy for it to tip over.

PREPARING THE WORK AREA

Before riprap can be placed, a suitable keyway foundation must first be dug, and the eroded slopes excavated to the designed slope. A keyway is a trench dug along the toe of the riprap fill, its purpose is to keep the riprap from being undermined by shifting sands and gravels. A typical riprap installation with keyway is shown in Figure 12-2. The depth of the keyway varies, usually from 2 to 4 feet deep, with the size of riprap to be placed.

One problem with riprap keys occurs when bedrock is encountered a foot or so above trench grade. It would be very expensive for a contractor to drill and shoot the rock. Moreover, there is no reason to go through this trouble. Bedrock, of course, is the most stable foundation for riprap. If it is not already in the contract, it is best to get the bedrock problem out of the way early by discussing it with the engineer. A work order should be written stating, "The contractor shall construct keyways to specified depth or bedrock, whichever comes first."

Keyways can be dug with a crane and clamshell or a hydraulic hoe. A crane is much slower than a hydraulic hoe, but has the advantage of much greater reach. For slopes that require a reach of 30 feet or less, hydraulic hoe is the machine to use, but for fill slopes more than 30 feet down, a crane is best. It is not unusual to have eroded areas a hundred feet or more below a road. A close-up of a clam bucket is shown in Figure 12-3.

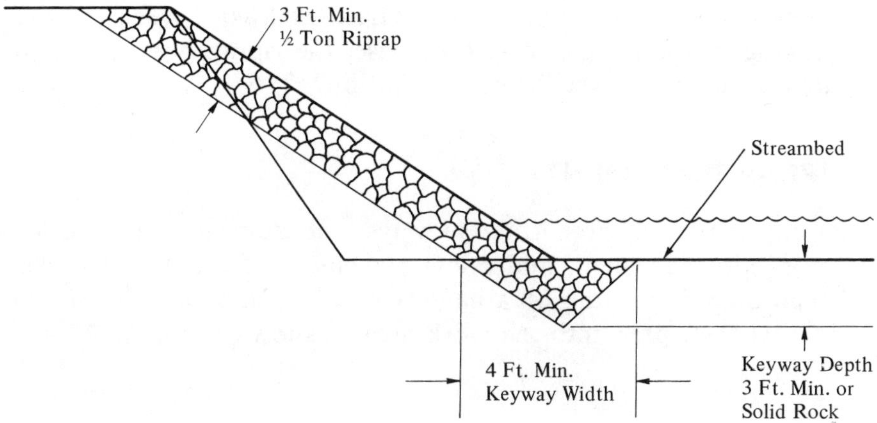

Fig. 12-2. Typical riprap illustration showing riprap thickness and keyway width and depth.

Fig. 12-3. A clam bucket on a crane is best for excavating the keyway when the work area is lower than 30 ft. below the machine.

Keyway excavation should be started at least 2 weeks before bringing in the rock trucks, otherwise, the riprap operation will catch up to the slower keyway operation.

RIPRAP PLACEMENT

Riprap is placed three different ways: end dumping with trucks, crane-skip placement, and crane placement of individual rocks. When dumping riprap on a narrow road, a small dozer is used to sidecast the riprap into the work area as shown in Figure 12-4.

Fig. 12-4. When dumping riprap on a narrow road, a small dozer is used to sidecast the riprap into the work area.

End Dumping With Trucks

This is the least expensive method of riprap placement. The loaded trucks back up to the edge of the eroded area and end dump the riprap into the keyways, which tend to catch the rolling rocks. End dumping is used when fill slopes are not too high (up to about 30 feet). One of the major problems with end dumping is that the large and smaller riprap tends to become segregated as it rolls down the fill slope. And since most contracts call for a well-graded mixture of rock, the crane or hoe may have to do quite a bit of the mixing.

Crane-Skip Placement

A skip is a lightweight dump truck bed or other boxed container rigged to a crane for placement of riprap at the bottom of long fill slopes. End dumping down a 100-foot slope causes segregation as with the 30-foot slope, and the impact at the keyway breaks up much of the riprap. The skip has a pair of heavy cables about 8 feet long tied at opposite corners of the box. The two cables on each side are then tied to one of the two cables on the crane. A truck backs up to the skip and dumps its load into it. The crane then lowers the skip down to the riprap area. When the skip is where the riprap should be dumped, the crane operator releases the brake on one of the two cables and the box tips to dump its load.

The skip placement method is admittedly more expensive to perform than end dumping directly into the work area, but it is not as slow as it sounds. Once an efficient routine is developed, a crane can place hundreds of tons of riprap each day. One of the big advantages of skip placement is that the "combing" crane does not have to mix the riprap because there is no segregation. Also, there is much less finishing work when smoothing out the riprap.

Crane Placement Of Individual Rocks

This is the most expensive and time-consuming type of riprap placement. It is used to place very large riprap (3 tons and up) in areas where proper placement and fitting of individual stones is critical to the stability of the riprap. Sometimes this riprap is put on very steep, and even vertical slopes. Extreme measures such as chipping away irregularities and squaring stones are sometimes employed to minimize voids between rocks. These rocks may also be drilled and a rock bolt installed so the crane will have something to hold on to. All of these extraordinary measures should be bid according to the amount of extra time and equipment involved.

RIPRAP COMBING

Once the riprap is placed in the eroded areas, it must be worked and shaped to minimize voids and irregularities. This is always a stipulation in a riprap contract. This working and shaping is commonly called *"combing."* Combing is very time consuming in end-dumped areas. The rock must be remixed, and rocks that have rolled past the keyway must be brought back in and "hid" in the riprap slope. If loose rock is called for, it is spread evenly over the

Fig. 12-5. One-ton size riprap in place and combed. Next, rock fill will be spewed over the rock to fill in the voids.

riprap to fill in the large rock voids. Figure 12-5 shows riprap in place and combed before application of loose rock.

Combing in skip-placed areas is less expensive than in end-dumped areas because of the lack of segregation, but the slope of the riprap must still be smoothed and rock points flattened. There is little or no combing needed with individually placed riprap because of the extra care in placing it.

13
Drilling and Blasting

The intent of this chapter is to explain the basics of drilling and blasting. It is not a crash course after which a contractor can go right out and start blasting any rock bluff to perfection. The use of explosives in the hands of inexperienced men is dangerous; it's not even all that safe in the hands of experienced users. Training courses are held periodically by explosives dealers for the beginner to gain practical experience. Also, a friend who is experienced in blasting can show you how to use it. These are the best sources to learn about explosives and their uses. Dynamite is too dangerous to be learned out of the pages of this book.

Drilling and blasting rock is an uncertain science at best. In order to perform the "perfect" blasting operation, drill holes must be properly spaced depending on the density of the rock and the explosive used, and the amount of explosive must be precisely calculated to break the rock to the required size. Unfortunately, it would take a team of physicists and geologists to determine these exact quantities, and no two teams would come up with the same answer.

But however uncertain the art of drilling and blasting, a contractor can still suitably perform this work, which is simply to break the rock into pieces small enough to be handled by the earthmoving equipment.

DRILLING

The area to be drilled for blasting should first be prepared by a dozer. A drilling machine needs a flat area from which to drill down, and this area should be free from all overburden (dirt) and loose rock, otherwise the drill holes will cave in. It is also advantageous for the dozer to rip as much rock as possible so the drill will be doing what it is supposed to, that is to drill hard rock.

The air track drill as shown in Figure 13-1, is one of the most common pieces of drilling equipment. It is comprised of an air compressor that drives a track-mounted drill, connected by hose lines. The compressor is stationary while the air track moves about from hole to hole. A jackhammer can be used for small ditch jobs, or

Fig. 13-1. An airtrack drill driven by compressed air is the standard drill rig for construction. Note that the compressor stays on the road below while the airtrack moves around in the drilling area, connected by air hose.

when rock need not be drilled more than 5 feet. For deeper holes, the air track is superior in both speed and economy.

Air-track drills come in various diameters, but the one most commonly used drills a 3 inch diameter hole. This diameter has been determined to be the most economical for construction projects. Drill shafts, or steels as they are called, come in 10-foot long sections and are easily added for deeper holes by reversing the drill. A two-way collar quickly tightens the adjoining steels.

An air-track drill can bore through 30 feet per hour, and deeper in softer rocks such as sandstone or shale. However, a drill loses its effectiveness if the drill holes get deeper than 30 feet because of bending of the drill shafts. If a 50-foot rock bluff is being blasted, it must be drilled and blasted twice. Remember, the worst enemy of any drilling operation is a dull drill bit. An air-operated grinder should be available at all times to sharpen the bits.

Hole patterns are determined by scientific analysis of the rock strength, the depth of the rock, and the explosive to be used. As mentioned earlier, no one seems to have any formula for determining the exact hole spacing for any one job. There are, however, a few rules of thumb that should be adequate for the type of blasting required for a road job.

1. Spacing of holes should be equal to or less than one-half the depth of the holes to be drilled. Example: If a rock bluff requires 20-foot deep drill holes, then the spacing between holes should be no greater than 10 feet.
2. Spacing of the burden (distance from the excavation plane to the rock face) should generally be 4 feet, with the hole pattern staggered as shown in the illustration on page 103.
3. Subdrilling, the distance drilled below finished grade, should always be at least 4 feet to insure breakage.

Always make sure that enough air is available for the drill to operate at maximum speed and torque. Too often a contractor will use a compressor that just meets the drill rating, but air loss from the top of the hammer and hose couplings rob the drill of power. As a result the drill is always running as if overloaded, overheated, and with only partial efficiency. If there is not enough air to blow the rock dust out of the hole, the drill bit will become stuck in it.

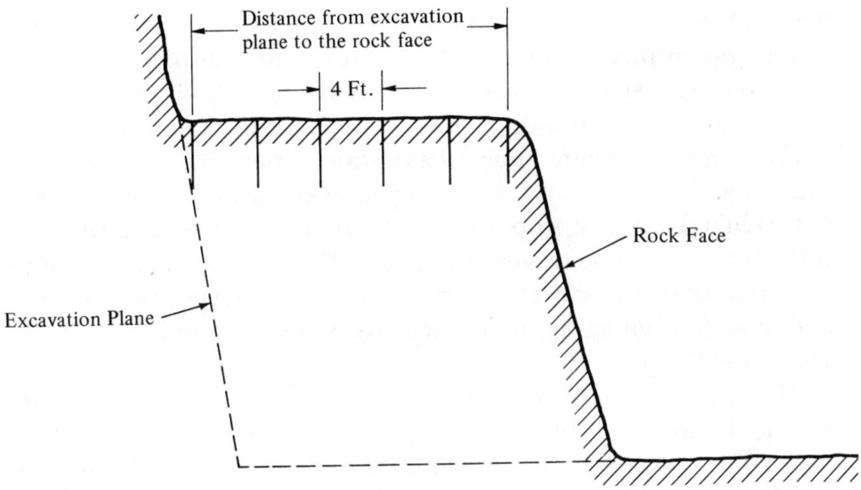

A drill should always be plumb and level. Too often drillers think that a drill can be leveled by eyeball. This is false. It is extremely difficult to do this accurately under any circumstances, and to do so without a level and plumb bob is nearly impossible. An error of a few degrees may mean several feet difference after the drill has bored 30 feet, and at the bottom where the explosives are packed, the holes may be almost touching. The result is pulverized rock in some areas and huge boulders in other areas which must be drilled and shot again.

EXPLOSIVES

The most generally accepted definition of an explosive is given as follows: "A solid or liquid substance or mixture of substances which, on the application of stimulus to a small part of the mass, is coverted in a very short increment of time to other more stable substances, largely gases, with the development of heat and high pressure."

An explosive device derives its disruptive force from the rapid conversion of its solid state to a gaseous state by burning. This burning occurs so rapidly that it exceeds the speed of sound, thus, the loud explosion. These gases are created in a confined drill hole and if unconfined, would occupy many times more volume than the solid explosive. The result is that extremely high pressures are exerted on the surrounding rock in the form of shock waves.

These high pressure shock waves are responsible for the actual breaking of the rock, but not in the manner that seems most apparent. While these waves do break some rock as they force outward, it is the return of these waves after they reflect off the open face of the rock that does the most damage. By then the waves are in disarray and interfere violently with each other, causing both shock and vibration.

The explanation of how an explosive works is essential in order to understand rock blasting. Since reflective waves actually cause rock to fracture, it becomes apparent that solid rock with only one free face is easier to blast successfully. Seamy rock with dirt layers in between reflects a free face at every seam. This is why seamy rock

Fig. 13-2. The result of an efficient blast—low noise and a humped-up, well fractured mass of rock that was once solid rock.

often is blasted in all directions with a thunderous roar, while hard rock simply humps out into a pile of broken rock with little noise. Figure 13-2 is the result of an efficient blast in what was once solid rock. In seamy rock, hole patterns and explosive charges are more of a guess than a science.

Selecting Dynamite

The strength of dynamite is rated by the weight percent of nitroglycerine that it contains. A stick of dynamite rated as 40% strength contains 40% nytroglycerin by weight. One misleading conception about the strength of dynamite is that the percentages are cumulative. For example, one might surmise that 60% dynamite is three times as powerful as 20% dynamite. This is not correct. Present day fillers such as sodium nitrate and carbonaceous fuel add substantially to the energy of the explosive. Consequently, 60% dynamite, which contains 60% nitroglycerine, is only about one and one half times as powerful as 20% dynamite.

Filling holes with high strength dynamite is not economical, and the results are often less than desired. Dynamite in these 60-80% strengths is used mostly in conjunction with a blasting cap to create a priming device which ignites lower strength explosives.

Blasting Gelatins

Blasting gelatin is a rubbery textured explosive packed in a sausage-like plastic tube (see Figure 13-3). These gelatins are becoming very popular because the plastic tube when slit lengthwise with a knife before it is dropped down a drill hole, will burst apart on impact and give the explosives a tighter mass from which to confine the gases. This is a critical advantage of gelatins. Unlike dynamite, gelatins are almost waterproof, making them much more suitable for blasting in wet bore holes, a constant problem for blasters. Toval is the gelatin explosive most commonly used in blasting for construction and quarry work. It is a very dense, economical gelatin varying in strength from 30-60%.

Fig. 13-3. Blasting gelatin is a mushy substance which has become very popular because of its capacity to form a dense mass in the drill holes. Gel is also waterproof.

Blasting Caps

Blasting caps are inch-long, silver-jacketed capsules of high explosives. Two wires extending from the blasting caps are wired to other caps to create the blasting circuit. When an electrical current enters the cap a heat-sensitive wire quickly heats the cap to the detonation level. The result is a chain reaction of cap to stick to gelatin. All this occurs in a few microseconds.

Priming

A blasting cap in conjunction with high strength dynamite creates a *priming device.* A wooden punch is inserted halfway into the length of a stick of dynamite and the cap is then gently pushed into the hole with the punch. This hole should not be so small as to cause the cap to be forced in, nor so large as to allow the cap to fall out.

When the cap is safely in place, a harness of wire from the cap is wrapped around the stick so that the cap will not fall out when the primer is lowered down into the hole. As a safety precaution, the primer should not be assembled until it is ready to be lowered into the drilled hole.

LOADING HOLES

When all holes are drilled and ready, the loading process begins. If the drill shows any signs of moisture, it is best to drop down a few gelatin tubes before dropping the water-sensitive priming dynamite. After each stick of explosive is dropped down the hole, a long tamping rod should be lowered to make sure the explosive did not hang up, thereby creating an air gap. Dynamite with a blasting cap should never be dropped down the hole; it should be lowered slowly to minimize the possibility of the blasting cap slipping out. When the proper amount of explosive is in the hole, it must then be *stemmed.*

Stemming is simply the process of covering up the dynamite so the valuable combustion gases cannot escape. Thin paper bags are

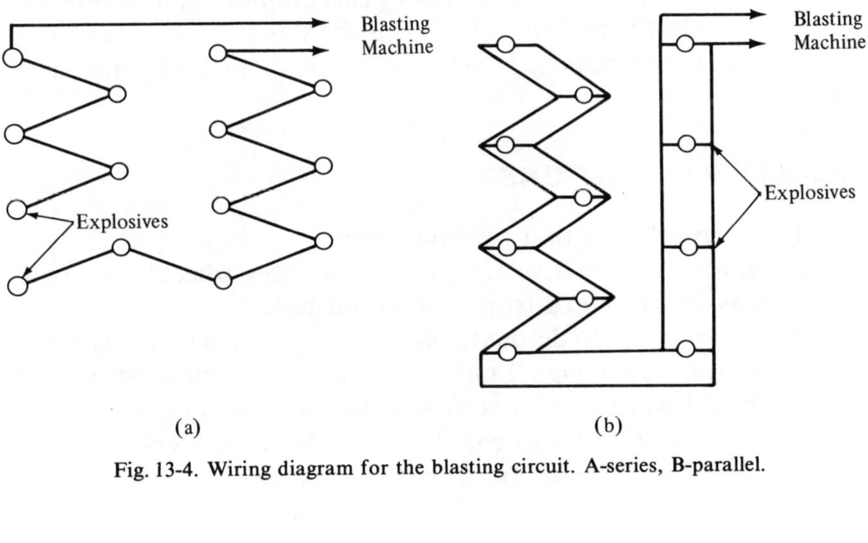

Fig. 13-4. Wiring diagram for the blasting circuit. A-series, B-parallel.

filled with sand or other fine dirt and dropped down the drill holes and then tamped with a rod. Tamping in long holes should be at least 6 feet deep.

The next step is the wiring of the blasting circuit. The series circuit (Figure 13-4A) is most commonly used and is the simplest way to wire explosives in up to 30 holes. For more than 30, a parallel circuit (figure 13-4B) is advisable to overcome any loss of electrical current and to ignite the caps. The series circuit is the safest method because the electrical continuity of the circuit can be checked with a galvanometer, an instrument that pushes a low, safe voltage through the blasting circuit. The test meter can show if the circuit is wired correctly and will detonate.

Detonating the Explosives

An electrical two-wire cord is reeled back away from the blasting area to a safe place, and the primer circuit is wired to one end of the cord. When personnel are clear of the area, the other end of the cord is wired to a blasting machine comprised of a dry cell battery, two hookup terminals, and a spring-loaded detonating switch.

The road should be blocked off from the time the primer circuit is hooked in. Once the blasting machine is wired in, detonation should be immediate. After the blast, wait for a few minutes before approaching the site. Toxic fumes of carbon monoxide and nitrogen oxides are formed from the explosives and can be harmful if inhaled. Severe headaches can also result from breathing these fumes.

SAFETY PRECAUTIONS

1. Always drill with a two-man crew.
2. Never drill close to fractured rock faces where loose rock may be dislodged from drill vibrations.
3. It is possible to detonate blasting caps with a two-way radio transmitter. A sign 2 miles on either side of the blasting area should warn motorists to shut off two-way radios.
4. When holes are loaded, all access should be blocked off for at least a half mile from the blasting site.

5. Never load explosives when an electrical storm approaches. If the blasting area is half loaded, leave it and get away.
6. Primers should only be handled by experienced personnel.
7. Never wire the detonating cord to the blasting machine first. The primer end should first be connected, and after workmen are clear of the area, the blasting machine can be wired in.
8. Drilling should be completed before loading begins. Never load holes when a drill is in operation.
9. Once the circuit is wired, detonation should be immediate. Otherwise impatient personnel may become anxious and leave their posts and see what is happening.
10. Store explosives in a safe, locked container where vandalism is least likely.
11. Personnel should move back at least one-quarter mile from the blast area and be ready to duck under a truck to avoid falling rocks.

14 Bridges

Nowhere in construction is so much work expended for such limited results. Several pieces of equipment and a dozen men may spend an entire season working on a bridge, and the end result is 100 feet of roadway over a stream, at a cost of $500,000. Bridge work, especially concrete bridges, is slower than road work, and more meticulous. When bidding a bridge job, a contractor should note each phase of construction, and decide how much it will cost in men, materials, and equipment. This is the only successful way to bid a bridge job. The work phases of bridge construction are:

clearing
structural excavation
foundation preparation
forms for concrete bridges
reinforcement bars
placement of the structure
finishing work, and
approach construction.

TIMBER BRIDGES

Timber bridges are usually single-lane structures averaging about 18 feet in width and 30 to 50 feet in length. They are generally more simple to construct than concrete bridges, and a contractor with sufficient manpower and knowhow can construct one in two weeks.

Such a bridge is shown in Figure 14-1, with the construction elements identified in Figure 14-2.

Timber bridges are almost always prefabricated at the plant to the plans and specifications of the contracting office. The usual procedure is that the contractor who is awarded a project is furnished with a set of bridge plans and specifications. It is his responsibility to obtain the structure. The contractor had previously submitted the plans and specs to the prefab bridge firm for pricing in order to prepare the bid. Having won the contract he now orders the bridge. A 3-month delivery time is standard. The engineering plans are made from a site survey, so a prefabricated bridge usually fits fairly well. The drawings note that the bridge will be prefabricated.

When the disassembled bridge arrives it should be checked to make sure that all parts are included and that the timbers are in good condition. Timbers that have been gouged or splintered will not be accepted by the engineer. A prefabricated bridge should also have all the hardware such as nuts and bolts, nails, plates, and anchor irons.

Fig. 14-1. A prefabricated, treated timber bridge complete except for guardrails.

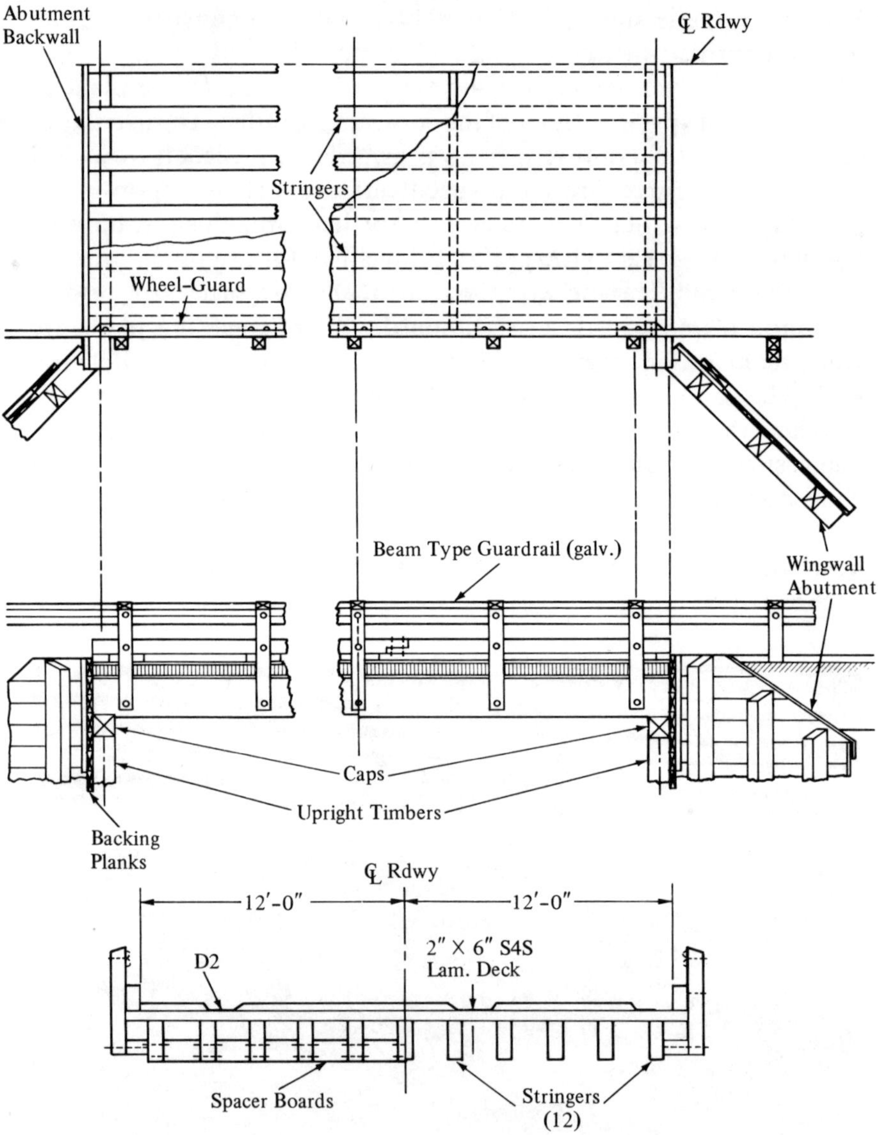

Fig. 14-2. Typical timber bridge terms.

An undeveloped bridge site is very confusing. The plans will call for a structure to be built 10 feet above a stream with riprap, guardrails, and approach road, but in the contractor's preliminary investigation the area may be nothing more than a brush-choked creek bottom. The best thing to do before starting any work is to consult the engineer and have him show you the exact site of the bridge, where it will span the creek, where the abutments and wingwalls will be, etc.

It is the engineer's responsibility to stake the bridge and reference each point so that it will be easy to reestablish after the clearing and excavation. Sometimes the engineer will reset those critical stakes, but he is not contractually bound to do this task, so the contractor may be stuck with the job. In this case, a level and transit will be needed. These instruments can be costly, and run into hundreds of dollars for those of a better quality. However, less expensive ones that cost a hundred dollars will suffice. These instruments can also be rented by the week from U-rent outfits. When clearing begins, care must be taken to avoid tearing the bridge stakes and reference points. This problem is especially critical in brushy creek bottoms. These stakes will guide the entire bridge construction.

Foundations

There are three types of timber bridge foundations: concrete, mudsill, and piling. The concrete and mudsill foundations will be described first, together with the rest of the bridge structure, before going into piling foundations.

Concrete Foundation

For a concrete foundation, a backhoe must excavate the abutment and wingwall areas to the designed grade or to bedrock, which is the most stable foundation for a bridge. This designed depth is the result of core drilling at the bridge site to find out at exactly what level hard-packed gravel can be expected, which is the minimum foundation for a bridge.

When hard pan gravel is encountered, the excavation will go much slower because of the compacted gravel and rocks that are just above bedrock. There should be some kind of agreement with the engineer that no more excavation below bedrock will be necessary unless a work order paying for drilling and blasting is issued. Usually blasting is unnecessary, but if bedrock is encountered only a foot or so below the streambed, it may be necessary to blast in order to provide stability for the foundation.

As soon as the excavation drops below the creek level, water will fill the excavated holes, and pumping will be necessary for the backhoe operator to see what he is doing. A 3-horsepower pump is adequate for timber-bridge foundations.

A concrete footing for the foundation may be required to be form-constructed, or it may be simply poured into the excavated hole to the required foundation grade. Either way, this is an easy foundation to construct, but remember to sink the rebar pins into the concrete as noted on the plans.

Mudsill Foundations

A mudsill foundation is a series of 18 × 24 inch timbers which lie flat on the foundation gravel or bedrock. It is sometimes tricky to keep the timbers level and free from warping. Gravel should be added to the low spots to stop the tilting since this does not hurt the foundation. The mudsill should have steel rebar and pins installed which fit into a drilled hole in the bottom of the upright timbers to keep the upright timbers from sliding during backfill tamping. A crane should be used to move the heavy timbers.

Once the mudsills are in place, they must be checked to see that they are parallel to each other and perfectly square. This is found by applying the following formula:

$$A = \sqrt{B^2 + C^2}$$

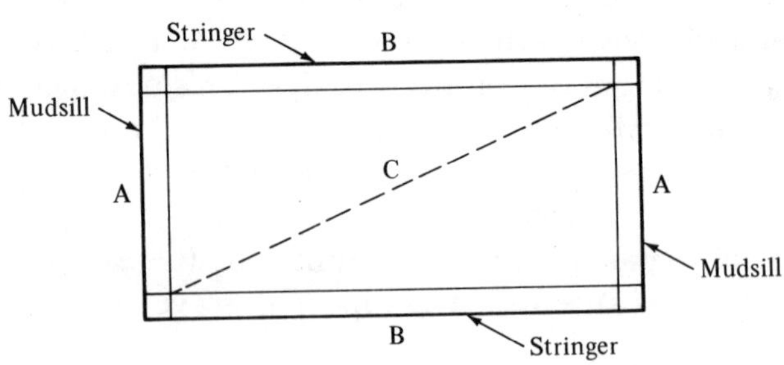

This is called the *Pythagorean* Theory by mathematicians, and simply called *figuring the diagonal* by construction people.

When the mudsill or concrete foundation is perfectly square and the rebar pins are in, the vertical timbers are placed on the pins. These are held upright by a few boards tacked to them and wedged into the dirt. When all the uprights on one side have been erected, the cap, or cross member is laid across their top. This too, is connected to the uprights by a drift pin rebar.

After both sides are up and capped, the two outside span stringers are placed on the caps, and the entire structure is squared up one last time. Up until now any part of the work could be redone, but once the stringers go on, you're stuck with it. The stringers are fastened to the caps and uprights by steel plates using 4-inch long screws. Spacers are also required between stringers. These are wooden blocks which keep the stringers from warping and buckling. Figure 14-3 shows timber stringers laid on the cap. Note the spacer boards and angle brackets.

The backing planks are assembled after the stringers. These are 2 × 12 inch planks which are nailed to the back of the abutment and wingwall uprights. Theoretically, the bridge is cut so that the last

Fig. 14-3. Timber stringers laid on the cap. Note the spacer boards and angle bracket which tie down the stringers.

backing plank fits perfectly to the top of the stringers, but that is never the case. The last plank must always be planed or a few inches cut off from its width to make it fit.

The Deck

There are two types of decks, laminated 2 × 6 inch planks or individually nailed 2 × 6 planks. The laminated deck comes in two pieces and installing it is a snap provided the bridge is perfectly square. However, small variations become big problems when attempting to fit two perfectly square pieces of deck onto a bridge frame that is not quite square. Nailing individual 2 × 6's together and to the stringers sounds like a lot of work, and it is compared to a preassembled laminated deck. However, when the bridge is not perfectly square, as is the case with most, it is much easier to make an individual 2 × 6 inch deck fit when the frame is not square. Figure 14-4 shows how 2 × 6 inch deck planks are nailed to the stringers.

Fig. 14-4. Laborer nails 2×6 in. boards on top of stringers to create a deck.

When the deck is on, the guardrail posts can be placed and the guardrail installed on the bridge, but not on the approach.

Backfilling can be done at the same time the bridge is being constructed. As soon as the uprights are placed on the foundation, the excavated trench should be backfilled and tamped with a hand tamper at least up to the level above the water. This will not only do away with the need for the pump, but will also hold the bridge frame rigid.

As soon as the backing planks are installed, the area behind them can be backfilled. Backfilling should be done simultaneously on both sides of the bridge, or one side should be compacted for about 1 foot, and then the other side filled and compacted a like amount. If one side is backfilled all the way to subgrade, it will cause pressure on the bridge and make the structure list away from the backfill. If this happens, the backfill will have to be completely removed and the bridge made true again.

Anchors

Anchor rods are essential to keep the earthen embankment from pushing out the abutments and wingwalls. Anchors are $1\frac{1}{2}$ inch diameter rods that are threaded at both ends. One end goes through the appropriate upright and is bolted loosely. The other end is bolted to a piece of anchor steel that is embedded into the backfill. When the backfilling is complete the loose bolts in the uprights are tightened.

The remaining approach guardrails are then installed with a backhoe. It sounds inefficient to have to backfill the approach and then excavate it again for the guardrail posts, but it is much easier digging a small hole when the correct postholes are located on the finished subgrade.

Most prefabricated bridges come surprisingly well cut, but often there will be pieces that are too short or too long, and boards from the local lumber yard will have to be substituted. The engineer will usually allow for this substitution as long as the new wood is treated with several applications of creosote. This will not take more than a day or two as the creosote is a fast penetrating product.

Piling Foundation

Piling foundations are used when there is no hard packed gravel or bedrock within several feet of the stream bottom. Pilings for timber bridges are round and look like telephone poles. The pile is tapered, with the diameter at the upper end larger than the lower end. This taper not only helps to sink them, but also gives structural strength because the larger diameter provides more bearing area. The large end of the piling has a heavy metal band attached to it so the end will not split from the force of the hammer blows when sinking it.

Pilings are sunk with a crane using either a diesel hammer, or a steel hammer and leads. The diesel hammer is the most accurate and efficient way to sink pilings. The diesel collar is fitted around a piling, which has been positioned, and the crane picks up the hammer and drops it once. The hammer hits the plate on top of the piling and drives the piling down a foot or so. This concussion sets off an explosion in a special compartment in the diesel hammer where a quantity of diesel fuel and igniter is fed. A small, controlled explosion results and forces the hammer back up to the height from which it was dropped. From then on it is a simple matter of the crane guiding the piling down at the correct spot into the creek bottom along a vertical line.

The hammer and leads method is slow and not nearly as accurate. A lead is a steel slide along which the hammer slides down to hit the top of the piling. The leads are tied to the piling and the entire unit is picked up by the crane and placed in the correct position. The hammer and leads method of sinking piles is shown in Figure 14-5. At this stage the piling is very wobbly thus reducing the chances of starting the pile vertically and in the correct spot. Very often the piling gets misaligned at the start and the crane cannot pull it back up. The only hope is an understanding engineer. Chances are that the piling will have to be dug up with a hoe, and the excavated trench backfilled and compacted before the piling can be tried again.

Pilings are trouble even in excellent soils. Add a few rocks or boulders to the soil and there is a likelihood that the pilings will be bent. A contractor is responsible for sinking vertical pilings, but not

Fig. 14-5. Piling foundation using hammer and leads is difficult because the piles do not go in vertically. Note the backhoe chained to the piling to correct a flaw.

if there is a hidden obstruction beneath the surface that makes them bend or slant. If the engineer insists on vertical pilings, compensation should be made for excavating the loose rocks and then backfilling the area.

Pilings which appear to be vertical when sunk are apt to be off a little when it comes to lay the cap across them. There will be small variations in the piles that are very difficult to straighten out when putting on the cap. Piles are as big as telephone poles and are not easy to move.

The best way to get all the pilings under the cap is to pin one at a time by pushing the bent pilings into place with a dozer or backhoe. More trouble may be encountered when laying the stringers because the bent cap may be difficult to keep square, but once the stringers are in place, the rest of the bridge structure is the same as for a foundation type bridge.

CONCRETE BRIDGES

Concrete bridges are difficult to bid because of the many small jobs connected with their construction such as removing the old bridge, setting up and removing concrete forms, and miscellaneous carpentry work. These tasks are difficult to estimate unless a contractor has had previous experience with concrete bridge work.

A concrete bridge with a poured foundation and deck does not go up overnight. Nor does it go up in a month. It takes several months to complete a bridge of even moderate size and complexity. So beware. The lucrative $500,000 bridge contract means months of slow and exacting work. Before any construction begins, the supervisor should visit the bridge site with the engineer to have him point out where the abutments, wingwalls, and piers (if any) are to be placed. Also, the stakes and references can be discussed and located.

Clearing is the first item of work and, as noted earlier in this chapter, care should be taken to avoid knocking out any stakes. Some contractors will drive reinforcement bars into the ground so that an errant tree will not wipe out the stake's location. The approaches to the bridge abutments are then developed and made suitable for a large hydraulic hoe to begin digging the foundation. Hopefully, bedrock will be encountered in the abutment and pier foundations near the foundation grade. If not, the contractor may be required to dig deeper. If bedrock is still not reached, steel pilings must be driven in clusters anywhere from 20 to 40 feet deep to support the bridge structure. A diesel hammer is the fastest and most efficient method of pile driving. Pilings are left about 6 feet above the ground so that the foundation concrete will distribute the load evenly to all pilings.

The foundation will be below stream level and some measure must be taken to keep out the water. A coffer dam is used around the foundation area for this purpose. A coffer dam is a series of interlocking pieces of heavy sheet steel which are lightly driven into the ground. It is not completely watertight, however, and a pump will be needed to remove the seepage. The engineer will require the foundation area to be free of most of the water so care should be taken to get the coffer dam as tight as possible.

When the concrete is poured and has set for a few hours, the coffer dam can be pulled out. No other work can be done on the fresh concrete until it has cured for the time specified in the contract, usually about 2 weeks. The elevation of the finished foundation will be just a little below the streambed grade. The foundation can be left in its rough state because the pier or abutment will rest upon it and the remainder will be covered by stream gravel.

The wooden forms for the piers (center supports) and abutments take much time and effort to construct. If the forming design becomes intricate, a contractor may have to hire a few carpenters for the more complicated work. Forming may take anywhere from 1 to 2 months. Rebar must also be bent and tied together to the configuration shown on the plans. This work can be done at the shop to ease the congestion at the work site. The cutting torches at the shop come in handy for heating the rebar to soften it for bending.

When the wooden frames for the piers are completed and greased, and the rebar cages installed, the concrete is poured. Once again, no work can be done on the concrete until the 2-week curing period is over. During this time odd jobs can be done, and arrangements made for the arrival and installation of the main bridge beams (stringers).

Most bridge stringers are prestressed concrete "I" beams ordered from the factory. If the bridge span is short, the stringers will run from one abutment to the other without a center pier. If the bridge has a span of over 100 feet, a center pier is needed, and the stringers will be shorter, running from each abutment to the center pier. Stringers never run between abutments in one piece with a center pier for support because of structural stress problems. Figure 14-6 shows prefabricated concrete stringers being laid between an abutment and center pier.

It is no simple matter to take a 90 foot long concrete and steel bridge stringer from a flatbed truck and lay it across the abutments. A stringer this long will weigh about 35 tons, and one crane will not be able to hoist it. Instead, one crane is stationed at each abutment, and one of the cranes slides the stringer off the truck and onto a heavy-duty conveyor belt. The stringer is slid along until it is parallel to the abutments, then each crane hooks onto an end and hoists

Fig. 14-6. Prefabricated concrete stringers being laid from one abutment to the center pier.

the stringer into place. Days, and even weeks, have been lost trying to get heavy stringers in place. The above method is the accepted procedure for experienced bridge contractors.

Once the stringers are in place, the deck forming can begin. Most contracts require the contractor to submit a deck forming plan for the engineer's inspection and approval. Many tons of rebar must also be bent and wired inside the bridge deck forms, and there is no way to do this beforehand. It must be done on the spot and can hold up deck pouring for weeks. A typical deck form showing the reinforcing bar pattern can be seen in Figure 14-7.

A very tight tolerance (about ⅛ inch in 20 feet) is used on the finished bridge deck, and only a paving machine can produce such results. The concrete is poured ahead of the paving machine through a 10-inch diameter hose, and the paving machine runs back and forth on rails along the length of the bridge to pave about two feet at a time. The paving machine can be calibrated to pave in the correct depth, camber, and crown. The paving machine makes two passes, first to skim off any excess concrete, and then to finish grade the concrete.

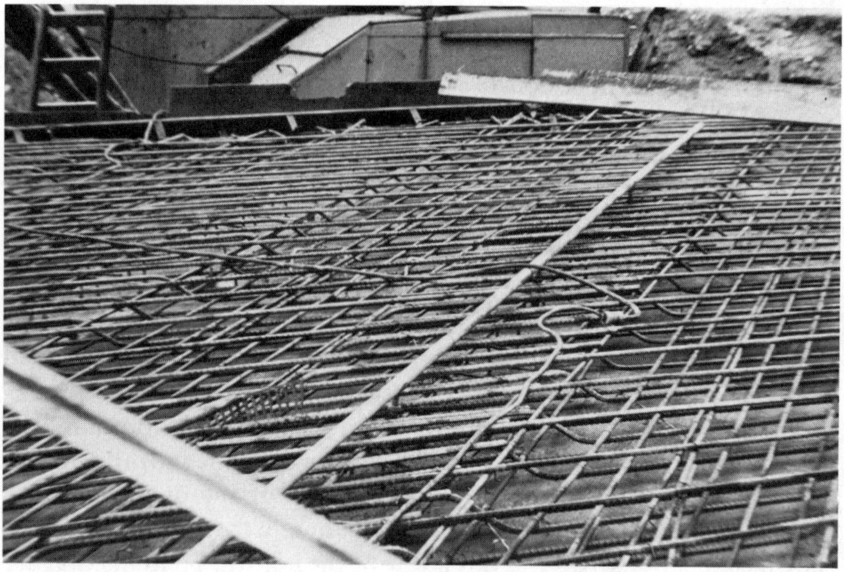

Fig. 14-7. The rebar forming for a concrete deck is slow work. There is no way to start the deck rebar cage beforehand.

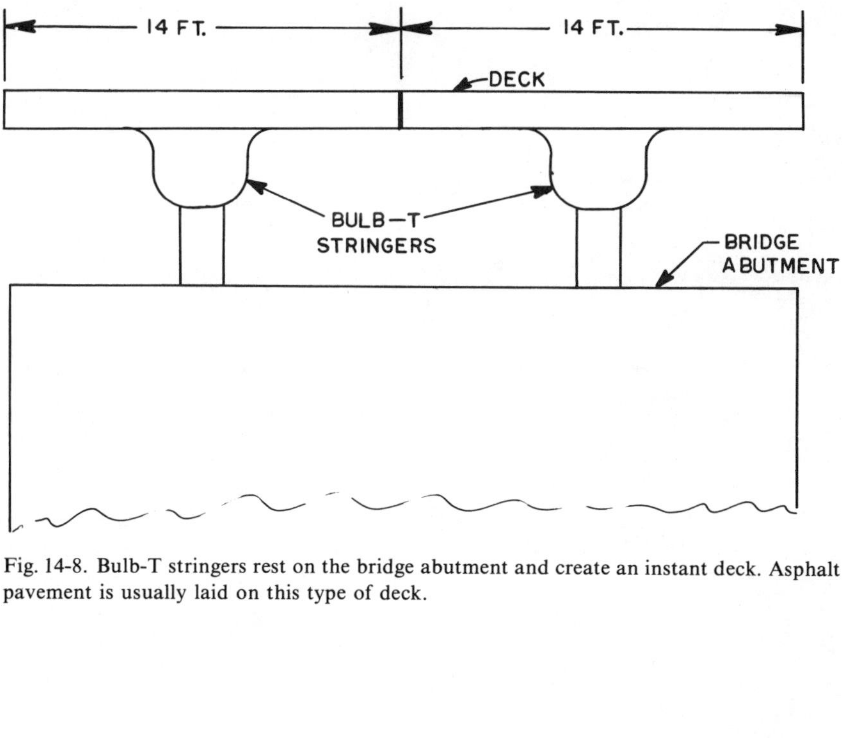

Fig. 14-8. Bulb-T stringers rest on the bridge abutment and create an instant deck. Asphalt pavement is usually laid on this type of deck.

The deck must also set for 2 weeks while the concrete cures. However, after the concrete is only an hour old and before it hardens, a steel or wooden-bristled broom is scraped across it. This leaves small, irregular grooves in the concrete which harden to give a skid resistant, rough surface on which to drive. If asphalt pavement is to be laid over the deck, the grooves give the asphalt something to grip.

On smaller span bridges of up to about 60 feet, a preformed bridge deck that is part of the stringer may be needed (see Figure 14-8). When the abutments are poured and cured, these "Bulb T" stringers are laid in place and an instant deck results. Preformed decks usually have asphalt pavement laid over them.

There are many other small items of work that accompany the construction of a bridge that have not been discussed for the sake of simplicity. Cold joints to allow expansion of the deck and cork liners which keep individual pieces of concrete from cracking when they move under stress are but a few of the additional items that are routinely encountered in bridge construction.

15 Grade Staking, Level and Transit Use

GRADE STAKING

Grade staking is used in construction to make sure cut and fill slopes and finished grades are built to the design requirements. Every contract has certain slope and grade tolerances to which the road must be built. Cut and fill slopes may be allowed to vary one-quarter foot in every 30 feet and road grade elevations by a few hundredths of a foot. These are tight specifications, and in order to keep from violating them, a stake hop, as he is called in the construction industry, must constantly check the cut and fill slopes, and the finished grade elevations.

There is another very selfish reason for checking the slopes. A contractor does not want to move a single extra yard since he is paid only for the designed quantity. If cut slopes are steeper than designed, besides being in trouble with the engineer, the contractor will have moved much more yardage than was required. On small 10-foot cuts this may not amount to much, but when cuts are 60 feet high, several hundred extra yards may be moved for no reason.

Fill slopes are just as bad or even worse. If the fill slope is not checked as it is built, it invariably will be steeper than designed. Being built steeper means that more fill was used and the excavation will be depleted before reaching the finished grade. The engineer will order the contractor to get the extra fill material by scraping the excess dirt off the steepened fill slope with a large hydraulic hoe—a slow, costly process. If the contractor requests that the engineer give him a borrow source to complete the fill, the engineer will stipulate on the work order that the contractor agrees to excavate the borrow for no extra compensation since it was his fault.

Stake hop equipment should begin with a good hand level for a man to look through in order to get a level line of sight as he works

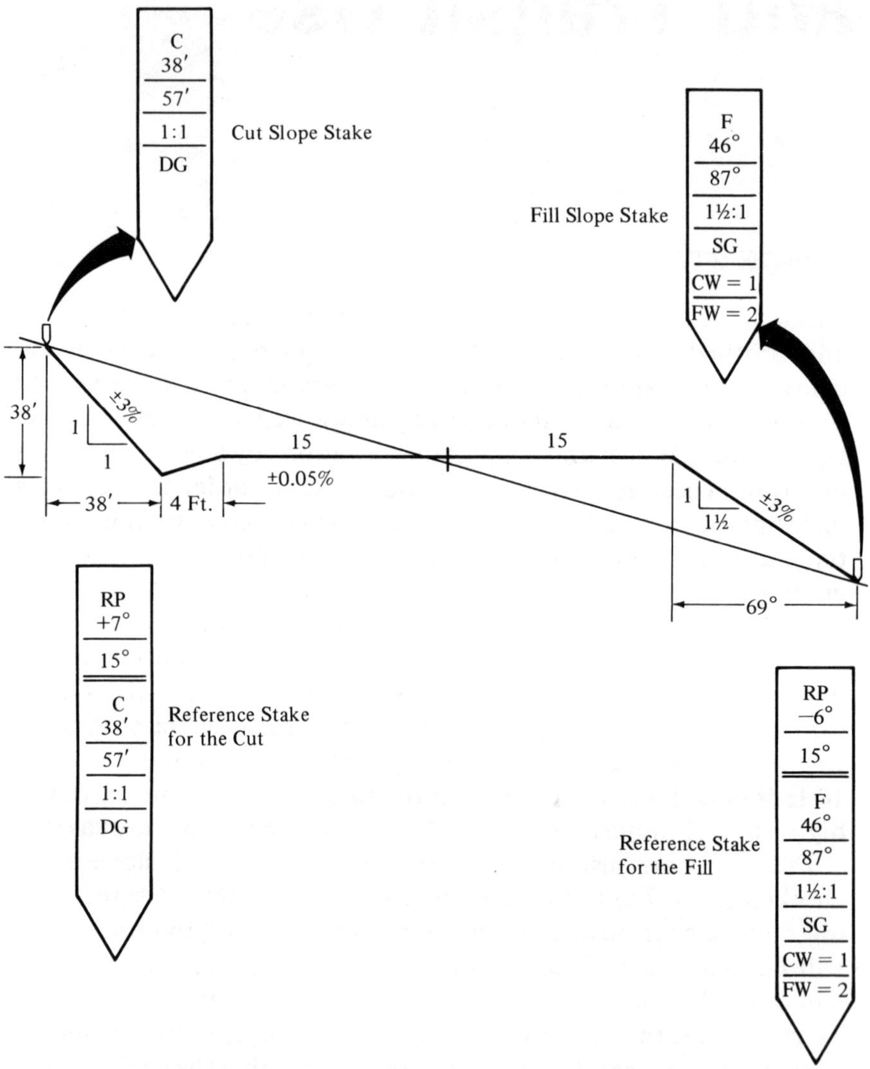

Fig. 15-1. Cut and fill slope stakes on a typical two-lane paved road.

down the slopes. A 7-foot measuring rod is also needed, together with a good supply of lath (30 inch long × ⅜ inch thick stakes used for marking variations in slopes and grades. The marking on the lath is made with a thick lead pencil or large felt-tipped marker. This equipment is available at any engineering supply store.

Figure 15-1 shows the cut and fill slope stakes and their tolerances on a typical two-lane paved road project. Variations will occur since, unfortunately, every state, federal, and private engineering department has its own way of writing stakes, but such variations are easily mastered. Before stake hopping begins, the engineer should explain how that particular road project was staked and what the writing on the stakes means.

Cut Slope Stake

The cut slope stake in Figure 15-1 shows the general information on such a stake. The "C" stands for cut. On this stake there is a 38.1 foot vertical cut to the hinge point, or bottom of the ditch. The next number, 57.1, is the horizontal distance in feet, from the cut stake to the centerline of the road. (All roads are staked horizontally with respect to centerline.) The nomenclature below that, 1 : 1 D.G., is the slope of the cut. D.G. indicates the cut is to the grade at the bottom of the ditch. A 1 : 1 slope means that for every vertical foot drop, you go out 1 foot horizontally. When the cut reaches ditch grade, the cut slope will be down 38.1 feet vertically, and 38.1 feet horizontally. The stake jumper on this particular slope will shoot the slope right after the cut is pioneered and put a lathe where the designed slope intersects the pioneer road as shown below.

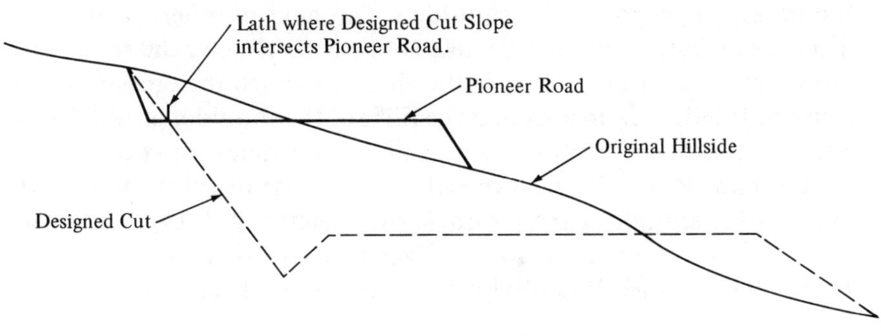

Fig. 15-2. Grade stakes every 6 ft down the cut slope tell the dozer operator that his slope is too steep and should be at the arrows on the lath.

He will check the slope every 8 feet or so as the dozer excavates the cut. That 3% tolerance on the slope means that on this 38.1 foot cut, the slope cannot vary more than 0.8 foot (3% of 38.1 feet). Figure 15-2 shows stakes on a cut slope that inform the dozer operator that his slope is too steep and should be at the arrow.

Fill Slope Stake

The fill slope stake shows that there is a 46.0 foot vertical fill up to the other hinge point, the shoulder. The next number, 87.0 is the distance in feet, from the fill stake to the centerline of the road. The numbers below that, 1½:1, is the slope to which the embankment must be filled. S.G. means that the fill is to the shoulder grade. CW = 1 means that there is one extra foot of curve widening on this section of the road. FW = 2 means that there are 2 feet of extra road width to allow for sloughing from rain, snow, or settling. A 1½:1 fill slope means that for every 1 foot of vertical rise there is a 1½ foot horizontal change. It is obvious that this is a flatter slope than the cut. This is necessary to stabilize the loose fill material. When the fill slope is to the shoulder grade, it will have come up 16 feet vertically, and over 24 feet horizontally. The stake jumper should establish the correct fill slope right after the fill is pioneered, and then check the slope about every 6 feet. A lath in the fill slope that was checked is all the equipment the operator needs to keep him on mark. If the slope is a half foot below the designed grade, the stake hopper should write −0.5. That will tell the operator that he has to steepen the slope a little. If the slope is one-half foot above the correct slope grade + 0.5 should be written, and the operator will know that the fill is too steep and that it will have to flatten out a little.

Reference Stakes

A dozer is a mammoth machine, and cannot be expected to miss every slope stake on the project. These construction stakes are constantly being knocked down by the dozers, especially when they are pioneering. Without some kind of reference, precious control would be lost every time a stake is knocked out. For that reason every fill and cut stake has a reference stake behind it. A short stake called a *hub*, is driven into the ground a set distance behind the cut and fill stakes, usually a constant of 15 or 20 feet. The vertical and horizontal distances from the construction stake to the hub are then measured and written under the R.P. so that a stake hop can relocate the cut and fill slope stakes where the stake used to be, should they be destroyed. The stake hop will also want to know the data that appeared on the original cut or fill slope stake, so this information is reproduced on the R.P. stake exactly as it appeared on the corresponding slope stake. Reference stakes are also shown on Figure 15-1.

The distance along a road project is marked by stations. Each station is 100 feet long. Every project begins at 0+00. A hundred feet into the job the station is 1+00 or 1 station. Fifteen hundred and forty three feet into the project, the station will read 15+43 stations. This stationing is written on the back of all cut, fill, and R.P. stakes. Project lengths are always noted in stations. For example, a road job that goes from 0+00 to 247+11 is 24,711 feet long. or 4.69 miles (5280 feet per mile). Station yards is a method of measuring haul. It is the product of the number of cubic yards of excavation multiplied by the distance in stations, or 100-feet increments. This is further explained in Chapter 18 under the heading entitled Overhaul.

If a project is small and has only one dozer, then that operator should be able to read construction stakes. If a project is large, with 20 machines, then all 20 operators should know how to read stakes, because sooner or later the stake hop will be busy somewhere else when an operator must find out about the slope or grade. If the operator can't read construction stakes he may ruin the area by proceeding without knowing what he is doing.

THE ENGINEERING LEVEL

A level is essential when constructing binwalls, timber or concrete bridges, or large pipes with concrete foundations. The principle of the level is simple. A horizontal line-of-sight is first established in the instrument. Using this, an operator can focus on an object to see if it is at the same elevation. All levels are set up on a tripod, a three-legged platform with adjustable legs.

Self-Leveling Level

A self-leveling instrument is the fastest and most accurate of all levels, and also the most expensive. To align this level there is a small water-filled sphere with an air bubble in it as shown in Figure 15-3. Three equally spaced thumb screws are used to get the bubble inside a painted circle in the middle of the sphere. When this is done, a small button is pushed or the instrument is tapped lightly with a finger, and a self-leveling prism working on gravity levels the instrument. However, before the prism can be operated, the bubble must be within the painted circle on the sphere. This type level takes no more than 10 or 15 seconds to set up and is preferred by most contractors.

Fig. 15-3. A self-leveling level is fast and easy to operate. Note the bubble within the circle, necessary to activate the internal gravity leveling prism.

Standard Level

The standard level must be aligned by adjusting four equally spaced thumbscrews until water-filled glass tubes have bubbles in the center of each tube. These tubes are at right angles to each other and the standard procedure is to level one, then turn it perpendicular and level the other. In doing this, the first bubble is knocked slightly out of adjustment, and the procedure must be repeated two or three times to overcome adjustment errors. After the instrument has been leveled, do not try to readjust it if one of the bubbles floats outside the middle of the tube. This is due to sunlight or cold expanding or contracting the fluid, and does not mean the instrument is out of adjustment.

TRANSIT

A transit is used for establishing precise angles and is essential for all bridge and concrete work. Figure 15-4 illustrates this important in-

Fig. 15-4. A transit is essential for all structural work.

strument. All transits come with a manual that explains the workings of the instrument, but much can be learned just from tinkering with it. A transit is set up on a tripod and must be exactly above the point of reference, be it a stake or nail. This is done by hanging a plumb bob from a hook on the underside of the instrument. A plumb bob is a 16 ounce carrot-shaped piece of brass that hangs at the end of a string. Gravity makes the plumb bob hang perpendicular to the earth, thus providing the perfect vertical alignment. The transit is leveled just as the standard level, by centering a bubble, turning the instrument perpendicular, and centering the other bubble. The bubble is shown centered in the leveling tube in Figure 15-5. Each bubble is leveled at least twice to compensate for adjustment errors.

All transit work is keyed to reference points. The instrument is set up on one reference stake in the ground and sights to another reference stake. This gives the exact line-of-sight required for that par-

Fig. 15-5. The bubble is centered in the leveling tube of the transit.

ticular job. If the location of a bridge abutment is to be reestablished, the transit will give the exact line-of-sight, and by measuring the correct distance from the reference stake at the transit, the abutment is accurately located. Backhoe excavation can then begin, but after the hole gets big and messy, the transit can give "line" to check if the machine has gone off course.

The vernier on a transit is a circular scale divided into 360 degrees and subdivided into minutes, which are parts of a degree. When the above mentioned abutment is located, the transit can then set up on that point, sight the reference stakes, turn the vernier 90 degrees, and measure the width of the abutment, thus locating it in two directions.

Transits are precise instruments that cost several hundred dollars for the better models. However, they can also be rented. Any time tolerances on a structure are given in hundredths of a foot, a level and transit should be used for this tight control.

16
Retaining Walls

Retaining walls are used on extremely steep or unstable sidehills where it is advantageous to build up vertically either the cut or fill slope. Retaining walls are also used where a road must run along a creek in the bottom of a steep canyon. In Figure 16-1 is shown a steel retaining wall to provide support for a road at the bottom of a steep gorge.

Retaining walls built into the fill slope have a history of being more stable than cut slope walls. The failure ratio is about 5:1 in favor of fill retaining walls. Whether a retaining wall is built into the cut or fill of a road makes a difference to the contractor because fill retaining walls are simpler to construct since all materials are hoisted down to the structure instead of up.

A retaining wall may be constructed using three types of material: concrete, steel, or timbers. No matter what type of material is used, a retaining wall is so simple a structure that any contractor with access to a hydraulic hoe can pick up a set of plans and proceed with the construction. A transit and level are essential because the elevations of the foundation and finished grade are critical. Also, the curvature of the retaining wall is needed to determine these points for the excavating machine.

CONCRETE RETAINING WALLS

Concrete retaining walls are used primarily on high cost freeway projects. The main advantage of this type of wall is that it is im-

Fig. 16-1. A steel binwall to provide a roadway at the bottom of a steep gorge.

pervious to rusting and rotting. Concrete retaining walls require a foundation to stabilize the massive weight of the concrete. This may be in the form of a 3-foot deep concrete footing along the base of the wall, or steel pilings driven into the ground with concrete poured around them.

Once the foundation is laid, it must be cured for about 2 weeks before work can commence. The wooden forms and rebar for the walls are not complex structurally and most workers with a knowledge of hand tools can accomplish the job. Concrete retaining walls require anchor rods which are embedded about 10 feet into the hillside and tied into the rebar. A backhoe is needed to dig a trench to lay these rods, and the trench is then compacted with a hand tamper. These rods keep the weight of the sidehills from pushing the wall outward, the major cause of concrete retaining wall failures.

STEEL BINWALLS

A steel retaining wall, or binwall, is constructed differently than the concrete type. Most important is the fact that a steel binwall needs no bedrock or concrete foundations to keep it in place. This type of structure rests right on the native soil or rock, and it is the mass of the fill material compacted in its bins that keeps the structure in place. Steel binwalls are purchased from a steel culvert company. After furnishing the factory with a set of plans and specifications, the finished structure will be available in about 3 months. This package will include all sheet steel, plates, connecting irons, and nuts and bolts.

Once the binwall excavation area has been staked, a hydraulic hoe can start excavation, but the volume of excavated material is usually excessive when working in a narrow area, and this dirt will have to be truck hauled to a waste site out of the working area. The transit and level should be used to keep the excavating machine on line and to the required depth.

Once the foundation elevation is reached, excavation should continue back into the hillside far enough so that workmen will have room to get behind the bins to tighten nuts and bolts. This foundation elevation may change for every bin, or it may remain constant if the topography is constant. When the elevations are constantly changing, the excavation to the foundation grade is much slower. Care should be taken during this operation to make sure that all grades are correct or else the binwall will be constantly binding and slow production.

The steel binwall comes in 8-foot sections. It is advisable to bolt about five bins together loosely at first because some holes are sure to be out of alignment. Pry bars will be needed to shift the steel sheets a little in order to get bolts through the holes. As the other sections are added, the first bins can be bolted down and filling and compacting begun. Figure 16-2 shows a steel binwall under construction. Sometimes two bolt holes will not line up no matter how much prying and bending is tried. This is common in steel structures with loose hole tolerances. When this happens the contractor

Fig. 16-2. A steel binwall should have several bins loosely bolted together before backfilling begins.

will have to drill new holes to get the bolts through. The engineer will allow this, but order the drill holes to be galvanized to prevent rusting. This is a simple matter for the contractor. A can of zinc chromate paint, available at any hardware store, is used to coat the holes and takes only seconds to do.

Fill for the bins is usually gravel or selected granular material that will drain moisture well. The use of gravel predominates. It should be stockpiled close to the binwall construction area so that quick access is available. This filling process goes very quickly, especially if there are two or three hand tampers in operation. It is enough to keep a backhoe running for more gravel fill all the time. In fact, another backhoe may be needed just to deliver the gravel from the stockpile to the work site, while the other backhoe dumps the gravel into the bins. The specifications for filling the bins usually call for 6-inch lifts to be compacted to 95% optimum density, which is easy to obtain with a hand tamper. The fills in the individual bins should not be brought up disproportionately. In other words, don't fill one bin to the top without filling any of the adjacent bins. This causes the same leaning and bulging as with culverts and bridges, and the contractor will be ordered to dig out the filled bin and straighten up the structure. However, the elevations of fill in adjacent bins can vary a few feet during filling without causing warping. A layer of riprap is often used around the base of the binwall along creek bottoms to keep flood waters from undermining the foundation. Binwalls are never vertical for structural reasons, and usually have a slight slope to them.

TIMBER BINWALLS

Timber binwalls are used mostly for esthetic purposes. Vacationers and tourists have found wood to be more pleasing along scenic

Fig. 16-3. The horizontal voids between timbers exposes coarse rock fill on a timber binwall.

routes. Timber binwalls are also much less expensive than the other types. They are manufactured by firms who make prefabricated timber bridges. Binwalls are made of 14-inch square timbers that are heavily creosoted to resist deterioration. The contractor must furnish the factory with a set of plans and specifications and then expect to wait 3 months before the structure arrives.

Timber binwalls are constructed much the same way as steel binwalls. The main difference is that the face of the timber binwall is not sealed. Figure 16-3 shows the horizontal openings between the timbers. For this reason select rock fill is usually required. This may be blasted rock or material from a rocky cut. Even with the rock fill, the specs will call for 1-foot lifts to be compacted with hand tampers. These machines do an excellent job of pounding down the rocks and knitting them together.

The only hardware used on this structure are the 1-inch rebar pins that tie the timbers together. There are no anchor rods or nuts and bolts with which to bother. Timber binwalls go up very fast, and two weeks may be adequate to completely construct a 150 foot long by 8 foot high binwall.

17
The Supervision and Orderly Operation of a Construction Project

The most lucrative contract in the world and the largest fleet of equipment do not assure a big profit. A profit is not assured at all unless the work can be performed economically. Most successful construction corporations display this fine-tuned preciseness of operation.

Much thought and planning goes into setting up an efficient operation. It is not the vice-president of the company who deals with charts and graphs, but the man at the jobsite who directs the day-to-day operations.

A project supervisor must plan his project well in advance so that any unforeseen problems will not cripple his operation. If, for example, the loader breaks down, a good supervisor will fall back on an earthwork operation for which the loader is not needed. Sure, the trucks will have to go home, but at least the dozers and other equipment can do some work. Another prime aim of the supervisor is to keep the various construction crews from stumbling over each other. If the culvert crew has the road dug up for several hours installing a culvert and the fuel truck cannot get by to fuel up the big machines, thousands of dollars will be lost. Every crew should know exactly where the fuel truck and mechanic's truck are at all times. These are emergency vehicles and they cannot be held up.

The supervisor, however, cannot be everywhere to direct simultaneous operations. He must rely on his foremen to direct the laborers and equipment operators according to his plan of attack. The stereotyped foreman on TV is a swashbuckling guy with cocked hard hat and abusive tongue who is ready to lick any man in his way. Only television could present such a gross misrepresentation. Any man who is responsible for work amounting to hundreds of thousands of dollars cannot be the fly-off-the-handle type. Yet many aspiring foremen believe that callous toughness and a big mouth are the only requirements for being a successful foreman. Here again, nothing could be further from the truth.

The construction foreman must be even-tempered enough to see a situation for what it is and react to it for the benefit of the project. Brawling with an inspector is not the way to do it. A foreman must have the respect of his men. Respect, of course, is something that is earned, not forced. Men without respect for their foreman are apt to sit down on the job as soon as he turns his head just to get even.

How does a foreman earn the respect of his men? There are several ways. Honesty above all is mandatory. If it looks like the crew might have to work Saturday to get a rush job done, don't wait until Friday to avoid the sour faces. Tell them immediately. They may not like it, but they will appreciate knowing early enough to cancel any plans. A foreman should get maximum production out of his men without being cheap. Get this production during working hours. If the work day is over at 5 o'clock, don't have the men work 15 minutes late every day. Besides being illegal, it really irritates the workers.

Controlled sternness is mandatory in a foreman. *Example:* On one road clearing project several years ago, a young foreman was invited to play cards with the boys one night and everyone played and drank late into the night. At eleven o'clock the next morning one of the sawyers who had played cards the night before said to the foreman, "I think I'll take a nap. I'm bushed after last night. I've cut enough trees to keep the others going for a couple of hours." The color drained out of the foreman's face. He looked the man right in the eye and said, "Well make it a good nap because it's the only one you'll get on this job." In no time the sawyer was back cutting trees with chips flying everywhere. He got the message loud and clear, and the foreman didn't have to swear a blue streak or fight him.

All men like to know why they are doing a certain job. This is especially true with equipment operators. The foreman should always explain why he is moving equipment from one place to another, or why operators should change their construction procedure. If the operators are not told why, they will begin to think that the foreman doesn't know what he is doing. It is not necessary to go into a detailed discussion. But if the earthwork operation is being moved 3 miles up the road because there is a large cut that can be excavated fast enough to make a big addition to the monthly payment, tell them exactly that.

DEALING WITH THE ENGINEER

Engineers, believe it or not, are human. They are people trying to do a good job, just like contractor, and they should be treated with the same basic respect due any man. The project engineer does, however, have a great influence over a project, and one of his decisions can cost a contractor thousands of dollars. For this reason, project engineers are not college kids fresh out of school. Instead, they must have many years of contract experience behind them before they reach the project engineer level.

Still, as noted above, engineers are only human. An engineer will be more receptive to a suggestion to save the contractor money if both parties are on good terms, and there have been no attempts to pass off shoddy work. The first time the engineer discovers shoddy work being purposely covered up, he will begin to look at *all* the work more closely with a cynical eye, and be more apt to pick a project to pieces. For example, a long, 2-foot deep thrufill is being built by trucks hauling excavation 3,000 feet. At 8 o'clock in the morning the roller breaks down. Immediately, the contract supervisor approaches the engineer and requests that he be allowed to continue filling since the 2 feet of fill will be compacted by the trucks and dozer, and a special effort will be made to have the dozer "walk the fill down." If the engineer is on good terms with the contractor, the chances are that he will agree that a 2-foot fill will not be lost because it was not rolled, and he will allow this exception with the stipulation that the roller be repaired immediately. If, however, the engineer and contractor are not on good terms, the engineer is apt to pull out his specifications and show the contractor where it calls for 10-inch fill lifts to be compacted by a roller, thus shutting down the entire operation. This is one of those grey areas where the engineer can go either way without hurting the project. Which way he goes will depend on how he feels about the contractor. A feeling that amounts to a mere $10,000 in lost time and production.

There will be times when the engineer and the project supervisor disagree. When thousands of dollars are at stake, bitter words may be exchanged. These things occur in the course of a contract. But once the argument is over, forget about it.

It is not a good idea for the project engineer or his inspectors to have direct relations with the contractor's laborers and equipment operators. If the engineer tells these men that he doesn't like the way some work is being done, the men may be confused for hours waiting for directions from their superior, when in fact the engineer was only suggesting a slight change in procedure. Also, some men get very irate when the engineer climbs onto their dozer and explains how the dozer should do the work. Every job has a pre-work meeting. Request that the project engineer and his inspectors consult only the project supervisor or his foremen.

PERFORMING REQUIRED WORK

A contractor is required to do all itemized and subsidiary work called for in the contract. It is the supervisor's job to see to it that no other work is done unless there is additional compensation.

For example, take the case of an old single lane road being reconstructed into a double-lane paved road. There are 20 metal culverts in the old road with a total length of 600 feet that the engineer wants the contractor to remove. The supervisor searches through the contract and finds no evidence of this work either as an item of pay or as a subsidiary item. The engineer flatly refuses to pay, insisting that it is standard procedure to remove old culverts. He orders the contractor to remove, squash, and then bury the old pipes in the fills.

The engineer is wrong. There is a specification for the removal of old culverts, and just because it was missed by the road designers does not mean that no compensation can be made for it. The contractor should go to the contracting office and talk to the contracting officer, the project engineer's superior. This will result in a written work order for the removal of the old pipes at a rate of $5 per lineal foot. This is $3000 that a quick thinking supervisor was able to get just by watching closely for extra work with no compensation.

Several of these extra items of work are hidden somewhere in every contract and may run into thousands of dollars. A good supervisor will always look for these extra work items. Some items commonly missed are obliteration of the old roadway, structural excavation for culverts, riprap for bridges, removal of old culverts, watering, rolling, slope rounding, and culvert inlet and outlet ditches.

18 Design Quantities

How designed quantities are obtained has always been a grey area for the contractor. He is constantly assured by the engineer that all quantities are correct since they came directly from the computer. Yet these designed quantities become very questionable when a contractor feels he is excavating or hauling much more material than the designed quantities call for. How then, does a contractor go about checking computer-designed quantities? The answer is, by manually computing the quantities in question and comparing them to the computer quantities. This chapter will explain how to calculate such major items of work as clearing, seeding, excavation, overhaul, and structural excavation.

THE COMPUTER

It has become customary in our everyday lives to avoid fighting the computer. Since we do not understand its space age technology and phenomenal ability to just about calculate everything, we begin to believe that the computer is a super electronic brain that thinks for itself and never errs. This is just not true. The computer is nothing more than a maze of electronic circuits, designed by humans, manufactured by humans, and most importantly—programmed by humans. When your monthly phone bill jumps from $15.00 to $16,351, the error is a human one—the computer only calculated what someone programmed. The same is true with computer road design; human errors go unnoticed and produce inaccurate quantities. The road designer fills out many computer forms to arrive at the final design quantities. One small digit in the wrong place can mean a loss to an unsuspecting contractor of hundreds or even thousands of dollars. That is why it is so critical for a contract supervisor and his foremen to be capable of manually computing work quantities. Every contract has design quantity errors, and it is up to the contractor to be able to find them. While some of the following calculations look complicated at first, with a little practice a contractor can run through any item of work in a few seconds using a pocket calculator.

PRELIMINARY DESIGN—THE SURVEY

The survey is the first phase of road design that directly affects the contractor because the designed quantities are only as accurate as the reference document. The location-line (commonly called L-line), which is the centerline of the proposed road, and the elevations of this L-line do not normally affect the contractor because they are surveyed with reasonable accuracy. However, the topography of the ground (called cross-sections), are used to determine most designed quantities, and this is the area where many inaccuracies occur. The problem is not that the surveyors take incorrect readings of the ground slope, but rather that the cross-sections do not accurately reflect the true ground slope. This happens when the cross-sections are not surveyed perpendicularly to the L-line of the road. This occurs especially around curves and on steep ground. Inaccurate cross sections are referred to as "skewed" cross sections and every job has a certain number of them. Some may be so slight as to make a negligible difference to the quantities, but on steep ground, cross sections which are skewed 10 degrees or more off perpendicular will make the ground appear less steep and give false information leading the contractor to believe that there are less quantities of excavation, seeding, and clearing.

After the award of the contract, but before the heavy construction begins, the contractor should request a copy of the road cross sections and compare them to the actual lay of the land. If faulty cross sections are discovered, the contracting office should be notified so that these stations can be re-cross sectioned and new quantities calculated.

EARTHWORK QUANTITIES

Excavation and embankment quantities are the most questionable items of work on a project because of the cross-section problem and because they comprise the most pay. Excavation and embankment quantities are calculated by multiplying the three dimensions of height, width, and length. The result will be in cubic yards, the standard of pay for excavation. Figure 18-1 shows cross sections at two station points with the proposed road in place. The first chore in figuring out the earthwork quantity within this stretch of road is to determine the distance between the two cross sections. As mentioned previously, the length of a road is measured by stations which are 100 feet long. Station 19+00 is 19 stations and station 19+50 is 19 stations plus 50 feet. Therefore the length of road between cross sections is 50 feet. We now have one of the desired three dimensions—length.

The simplest way to determine the next two dimensions, height and width, is to break the cut and fill into a series of right triangles (A thru E). For this, the contractor makes a drawing to scale, divides each cut and fill into right triangles and measures the height and width of each. This is the easiest way to determine the area of the cut and fill (commonly called the end area by engineers). The area of a right triangle is: $A = \frac{1}{2}HW$ (area equals one half the height times the width). The areas of all the triangles for each cross section can then be calculated and added together to obtain the area in square feet. Since the cut and fill areas will be different for each cross section, they must be averaged. This is done by simply adding the two end areas and dividing by two. In Figure 18-1,

$$\frac{60 \text{ square feet} + 92 \text{ square feet}}{2} = 76 \text{ square feet, average.}$$

This average is then multiplied by the length (50 feet) to give a total volume of excavation between the two cross sections of 1500 cubic

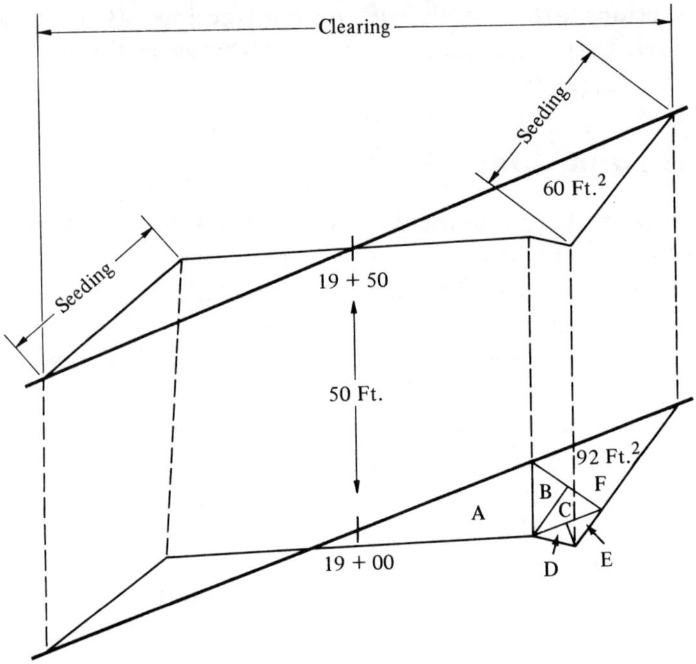

Fig. 18-1. Illustration of earthwork, clearing, and seeding measurements on a typical stretch of road.

feet. This is not the standard pay measurement so it must be converted into cubic yards. There are 27 cubic feet per cubic yard, so

$$\frac{\text{1500 cubic feet}}{27} = \text{55.6 cubic yards of excavation between stations 19+00 and 19+50.}$$

Structural excavation for culverts, retaining walls, riprap, and bridges is calculated in the same three-dimensional manner, but the actual calculating is much simpler since the excavation is in the form of a trench in which:

$$\frac{\text{length} \times \text{width} \times \text{depth}}{27} = \text{cubic yards of excavation.}$$

CLEARING QUANTITIES

Clearing is a two-dimensional quantity—width and length. In Figure 18-1 the clearing is calculated first by determining the length, which was previously found to be 50 feet. The other dimension, width, is measured horizontally from the upper clear limit above the cut to the lower clear limit below the fill. If the total clearing width at station 19+00 is 62 feet and at 19+50 it is 50 feet, then the average width would be

$$\frac{62+50}{2} = 56 \text{ feet, average.}$$

Since length multiplied by width gives the clearing area, 50 ft × 56 ft = 2800 sq. ft. However, the standard rate of pay for clearing is the acre, so divide the answer by 43,560 (square feet in an acre)

$$\frac{2800 \text{ square feet}}{43{,}560} = 0.06 \text{ acres of clearing between stations 19+00 and 19+50.}$$

SEEDING AND FERTILIZING QUANTITIES

Seeding and fertilizing are subcontracted to a landscape firm, so a contractor does not have to bother with this work item on a road project. However, the subcontractor may come to the contractor complaining that the number of acres being seeded far exceeds the designed quantities. In this situation the contractor will want to compute some of the seeding areas to determine the accuracy of the design quantities.

Seeding is a two-dimensional quantity like clearing, but differs in that it is not measured horizontally, but on the slope of the cuts and fills. Also, seeding is not calculated for the roadbed since that will have surfacing on it. Normally, the seeding for the cut side runs from top to bottom of the ditch, but sometimes the ditch slope up to the shoulder is also seeded. On the fill side, seeding extends from the bottom of the fill to the shoulder. These two slope widths are measured and totaled for each cross section, and then they are averaged just like clearing. The format for this calculation is

$$\frac{\text{average width, in feet} \times \text{length, in feet}}{43{,}560} = \text{acres of seeding.}$$

OVERHAUL

A dozer can economically move excavated material up to 300 feet before its efficiency drops off. That is why overhaul is used—to compensate the contractor for the extra effort it takes to move material over longer distances. If the overhaul limit is set at 300 feet, every cubic yard of excavation that is hauled a longer distance is paid for at the overhaul rate.

Overhaul is paid by the station-yard. If the pay rate per station yard is 10¢, the contractor is paid 10¢ per yard for every station (100 feet) that this cubic yard is moved past 300 feet. If a contractor must move 1500 yards of cut material 1200 feet, he will calculate the overhaul as follows: First, subtract the 300-foot freehaul distance from the 1200 foot haul, leaving 900 feet. Overhaul then is 1500 yards times nine stations, or 13,500 station yards. It is apparent that large quantities of excavation hauled thousands of feet can run into millions of station yards. That is why overhaul in great quantities is paid by the mile-yard which is defined as the movement of one cubic yard of material for one mile. The rate of pay for a yard mile is about $1.00.

Index